AF407849

Current and Future Developments in Proteomics

(Volume 1)

Microbial Proteomics: Development in Technologies and Applications

Edited by

Divakar Sharma

Kusuma School of Biological Sciences
Indian Institute of Technology, Delhi
India

Current and Future Developments in Proteomics

Volume # 1

Microbial Proteomics: Development in Technologies and Applications

Editor: Divakar Sharma

ISBN (Online): 978-981-14-9141-2

ISBN (Print): 978-981-14-9142-9

ISBN (Paperback): 978-981-14-9140-5

© 2020, Bentham Books imprint.

Published by Bentham Science Publishers Pte. Ltd. Singapore. All Rights Reserved.

BENTHAM SCIENCE PUBLISHERS LTD.
End User License Agreement (for non-institutional, personal use)

This is an agreement between you and Bentham Science Publishers Ltd. Please read this License Agreement carefully before using the ebook/echapter/ejournal (**"Work"**). Your use of the Work constitutes your agreement to the terms and conditions set forth in this License Agreement. If you do not agree to these terms and conditions then you should not use the Work.

Bentham Science Publishers agrees to grant you a non-exclusive, non-transferable limited license to use the Work subject to and in accordance with the following terms and conditions. This License Agreement is for non-library, personal use only. For a library / institutional / multi user license in respect of the Work, please contact: permission@benthamscience.net.

Usage Rules:

1. All rights reserved: The Work is the subject of copyright and Bentham Science Publishers either owns the Work (and the copyright in it) or is licensed to distribute the Work. You shall not copy, reproduce, modify, remove, delete, augment, add to, publish, transmit, sell, resell, create derivative works from, or in any way exploit the Work or make the Work available for others to do any of the same, in any form or by any means, in whole or in part, in each case without the prior written permission of Bentham Science Publishers, unless stated otherwise in this License Agreement.
2. You may download a copy of the Work on one occasion to one personal computer (including tablet, laptop, desktop, or other such devices). You may make one back-up copy of the Work to avoid losing it.
3. The unauthorised use or distribution of copyrighted or other proprietary content is illegal and could subject you to liability for substantial money damages. You will be liable for any damage resulting from your misuse of the Work or any violation of this License Agreement, including any infringement by you of copyrights or proprietary rights.

Disclaimer:

Bentham Science Publishers does not guarantee that the information in the Work is error-free, or warrant that it will meet your requirements or that access to the Work will be uninterrupted or error-free. The Work is provided "as is" without warranty of any kind, either express or implied or statutory, including, without limitation, implied warranties of merchantability and fitness for a particular purpose. The entire risk as to the results and performance of the Work is assumed by you. No responsibility is assumed by Bentham Science Publishers, its staff, editors and/or authors for any injury and/or damage to persons or property as a matter of products liability, negligence or otherwise, or from any use or operation of any methods, products instruction, advertisements or ideas contained in the Work.

Limitation of Liability:

In no event will Bentham Science Publishers, its staff, editors and/or authors, be liable for any damages, including, without limitation, special, incidental and/or consequential damages and/or damages for lost data and/or profits arising out of (whether directly or indirectly) the use or inability to use the Work. The entire liability of Bentham Science Publishers shall be limited to the amount actually paid by you for the Work.

General:

1. Any dispute or claim arising out of or in connection with this License Agreement or the Work (including non-contractual disputes or claims) will be governed by and construed in accordance with the laws of Singapore. Each party agrees that the courts of the state of Singapore shall have exclusive jurisdiction to settle any dispute or claim arising out of or in connection with this License Agreement or the Work (including non-contractual disputes or claims).
2. Your rights under this License Agreement will automatically terminate without notice and without the

need for a court order if at any point you breach any terms of this License Agreement. In no event will any delay or failure by Bentham Science Publishers in enforcing your compliance with this License Agreement constitute a waiver of any of its rights.

3. You acknowledge that you have read this License Agreement, and agree to be bound by its terms and conditions. To the extent that any other terms and conditions presented on any website of Bentham Science Publishers conflict with, or are inconsistent with, the terms and conditions set out in this License Agreement, you acknowledge that the terms and conditions set out in this License Agreement shall prevail.

Bentham Science Publishers Pte. Ltd.
80 Robinson Road #02-00
Singapore 068898
Singapore
Email: subscriptions@benthamscience.net

CONTENTS

FOREWORD i

PREFACE ii

ABOUT THE EDITOR iii

DEDICATION iv

LIST OF CONTRIBUTORS v

CHAPTER 1 MICROBIAL BIOFILM AND DRUG RESISTANCE: A PROTEOMIC APPROACH 1

Sarika Sharma and *Sandeep Sharma*

INTRODUCTION 1
   The Process of Biofilm Formation 2
   Attachment 3
   Micro-Colony Formation 3
   Detachment 3
   Mechanism of Biofilm Resistance against Antimicrobials 3
   Limited Penetration of Drugs Due to Glycocalyx 4
   Enzyme-Mediated Resistance 4
   State of Metabolism and Rate of Growth in Biofilms 5
   Genetic Adaptations 5
   Efflux Pumps 5
   Quorum Sensing 6
TARGET PROTEINS TO STUDY BIOFILM 6
   Proteins Involved in Cellular Metabolism 6
   Transporter Proteins 7
   Stress-responsive Proteins 7
   Proteomic Techniques to Study Biofilm 7
   Stable Isotope Labeling by Amino Acids in Cell Culture (SILAC) 8
   Isotope-coded Affinity Tags (ICAT) 8
   Isobaric Tags for Relative and Absolute Quantification (iTRAQ) 9
   Tandem Mass Tags (TMT) 9
   Isotope-coded Protein Label (ICPL) 9
   Bioorthogonal Noncanonical Amino Acid Tagging (BONCAT) 10
   Future Prospectus 10
CONCLUSION 10
CONSENT FOR PUBLICATION 11
CONFLICT OF INTEREST 11
ACKNOWLEDGEMENTS 11
REFERENCES 11

CHAPTER 2 PAST, PRESENT, AND FUTURE OF GEL-BASED MICROBIAL PROTEOMICS 18

Munna Lal Yadav, Arvind K. Verma, Preeti Rawat, Abhishek Parashar, Divakar Sharma, Sudarshan Kumar and *Ashok K. Mohanty*

INTRODUCTION 19
PROTEIN SAMPLE PREPARATION FROM MICROBIAL SOURCE 19
CLEAN-UP AND QUANTIFICATION OF PROTEINS 20
GEL ELECTROPHORESIS OF PROTEINS 20

TYPES OF PAGE 21
   SDS-PAGE (Sodium Dodecyl Sulphate-polyacrylamide Gel Electrophoresis) 21
   Tricine SDS-PAGE 22
   Native Page 23
   SDD-AGE (Semi Denaturing Detergent-agarose Gel Electrophoresis) 23
   2D-PAGE (Two-dimensional Gel Electrophoresis) 23
      Isoelectric Focusing (First Dimension) 24
      SDS-PAGE (Second Dimension) 25
2D-DIGE (TWO DIMENSIONAL DIFFERENCE IN-GEL ELECTROPHORESIS) 26
ANALYSIS OF 2D GEL AND DIGE GEL BY DIFFERENT SOFTWARE 27
   2D Gel Analysis using Image Master 2D Platinum 28
   Evaluation of DIGE Gels using DeCyder 29
COMPARISON BETWEEN IEF, SDS-PAGE, 2D-PAGE, AND DIGE 30
MICROBIAL PROTEOMICS IN DAIRY PROCESSING 31
FUTURE PERSPECTIVES OF GEL BASED MICROBIAL PROTEOMICS 31
CONSENT FOR PUBLICATION 32
CONFLICT OF INTEREST 32
ACKNOWLEDGEMENTS 32
REFERENCES 32

**CHAPTER 3 LIQUID CHROMATOGRAPHY-MASS SPECTROMETRY (LCMS): AN
          ADVANCED TOOL FOR THE MICROBIAL PROTEOMICS ANALYSIS** 36
Pranav Kumar Prabhakar
INTRODUCTION 36
MASS SPECTROSCOPY (MS) 40
   Shotgun MS-Based Proteomics 42
   Shotgun MS-Based Quantitative Proteomics 43
EXPERIMENTAL PROCEDURE 45
   i. Sample Preparation 46
   ii. Mass Spectrometry 46
DATA ACCOMPLISHMENT 49
PROTEIN IDENTIFICATION 51
**USE OF MALDI-TOF IN MICROBIAL IDENTIFICATION AND ANTIMICROBIAL
RESISTANCE** 52
CONCLUSION 52
CONSENT FOR PUBLICATION 53
CONFLICT OF INTEREST 53
ACKNOWLEDGEMENTS 53
REFERENCES 53

**CHAPTER 4 FUNCTIONAL ANNOTATION AND ENRICHMENT OF MICROBIAL
          PROTEINS USING SYSTEMS BIOLOGY: TOOLS AND APPLICATIONS** 61
Aditya Arya and *Vivek Dhar Dwivedi*
INTRODUCTION 62
IMPORTANCE OF ANNOTATION AND ENRICHMENT 63
EMERGING TRENDS IN GENOMICS 63
EMERGING TRENDS IN PROTEOMICS 64
ANNOTATION OF MICROBIAL PROTEINS 66
SYSTEMS BIOLOGY AND ANNOTATIONS 67
FUTURE GUIDELINES 73
CONSENT FOR PUBLICATION 73
CONFLICT OF INTEREST 73

ACKNOWLEDGEMENTS 73
REFERENCES 73

CHAPTER 5 EMERGING PARADIGM OF POST-TRANSLATIONAL MODIFICSATIONS IN MICROBES: AN UNDEFEATABLE WEAPON 76
Arpana Sharma, Chandrajeet Singh, Gopika Raval, Kruti Dave, Ankita Mathur
Juhi Sharma and *Divakar Sharma*
INTRODUCTION 77
    What are the General Characteristics of Bacteria and the Significance of the Study? 78
    What are the Post-translation Modifications? 80
    The Biological Significance of Post Translation Modifications and their Implications 83
    PTM Stories Unfolded in Some Pathogenic Organisms 86
    Role of PTMs in Host-pathogen Interactions 88
    PTM Role in Drug Resistance 92
    Mechanisms of Bacterial Drug Resistance 93
    Different PTMs Present in Bacteria 94
    Acetylation 98
    Phosphorylation 98
    Hydroxylation 100
    Lipidation 101
    AMPylation 101
    ADP-Ribosylation 101
    Glycosylation 102
    Carboxylation 102
    Nitrosylation 102
    Protein Pupylation and ubiquitin- like Modifications 103
    Proteomic Approaches Used in Exploration of Microbial PTMs 103
    Technical Challenges in the PTM Which Needs Immediate Attention 105
    The Lessons Learned from the Past and Giving Way to Future Solutions 107
CONSENT FOR PUBLICATION 107
CONFLICT OF INTEREST 108
ACKNOWLEDGEMENTS 108
REFERENCES 108

CHAPTER 6 PUPYLATION: A NOVEL PROTEOLYSIS PATHWAY IN PROKARYOTES FUNCTIONALLY REMINISCENT TO EUKARYOTIC UBIQUITINATION 122
Yogesh K. Dhuriya and *Divakar Sharma*
INTRODUCTION 122
    Bacterial Proteasome: An Evolutionary Precursor of Eukaryotic Proteasome 123
    Structure – Mycobacterium Proteasome 124
    Mpa –Mycobacterium Proteasome ATPase 125
    Pupylation –Mark of Intrinsic Protein Demolition 126
    Reversed Pupylation –Depupylation 127
    Recycling of Pup – Proteasome Regulation 128
    Pup-proteasome System (PPS) – Bacterial Physiology 129
    Proteasome – Stress Sensor Machine 130
    Cpa (Cdc48-like protein of actinobacteria) –A Novel Proteasome Interacting Molecule 131
CONCLUSION AND FUTURE PROSPECTS 131
CONSENT FOR PUBLICATION 132
CONFLICT OF INTEREST 132
ACKNOWLEDGEMENTS 132
REFERENCES 132

CHAPTER 7 OVERVIEW AND CHALLENGES IN PROTEINS SAMPLE PREPARATION FOR 2-DGE IN BACTERIAL PROTEOMICS 138

Divakar Sharma, Juhi Sharma, Nirmala Deo and *Deepa Bisht*

INTRODUCTION 138

Overview, Challenges and Probable Solutions 139

Enrichment and Extraction of the Complete Proteome from Bacteria 141

Protein Precipitation Strategies 142

SDS-Tri Chloroacetic Acid (TCA)-Acetone Precipitation for Cell Lysate and Cytosolic Proteins 142

Acetone Precipitation for Cell Lysate and Cytosolic Proteins 142

Ammonium Sulfate Precipitation for Secretory/Culture Filtrate Proteins 143

SDS-Tri Chloroacetic Acid (TCA)-Acetone Precipitation for Secretory/Culture Filtrate Proteins 143

Isolation and Fractionation of Cell Membranes Proteins through Ultracentrifugation 143

Effective Enrichment and Fractionation of Lipophilic Proteins without Ultracentrifugation 144

CONCLUSION AND FUTURE PROSPECTS 144

CONSENT FOR PUBLICATION 144

CONFLICT OF INTEREST 145

ACKNOWLEDGEMENTS 145

REFERENCES 145

CHAPTER 8 PROTEOMICS BASED BIOMARKER DEVELOPMENT AGAINST INFECTIONS: AN OVERVIEW 147

M. Madhan Kumar, Vivek Kumar Gupta and *Divakar Sharma*

INTRODUCTION 147

Proteins and Their Role in Infectious Diseases 147

Life is the Mode of Action of Proteins - Friedrich Engels 147

Techniques Used for Studying Protein-protein Interactions (PPI) 149

In Silico Methods for Studying Protein-protein Interactions 153

A Note on Biomarkers and Their Role in Infectious Diseases 153

Challenges in Studying Protein-protein Interactions 153

Protein Markers in Bacterial and Viral Infections – Pathogen and Host Perspectives 154

Influenza 154

Hepatitis C 155

Dengue 156

Human Immunodeficiency Virus -1 (HIV-1) 156

Severe Acute Respiratory Syndrome Coronavirus 2 (SARS-CoV-2) 157

Search for Protein Markers in Bacterial Infections – Host and Pathogen Perspectives 158

Use of Bacterial Markers for Typing Bacteria 159

Determining Antibiotic Resistance 159

Proteomics in Biomarker Discovery and Diagnostics 160

Biomarkers in Diagnosis, Treatment and Prevention of Infectious Diseases 161

THE WAY AHEAD – FUTURE PERSPECTIVES 161

CONSENT FOR PUBLICATION 162

CONFLICT OF INTEREST 162

ACKNOWLEDGEMENTS 162

REFERENCES 162

CHAPTER 9 MICROBIAL METALLOPROTEOME: APPROACHES AND BIOMEDICAL APPLICATION IN MICROBIAL ANTIBIOTICS RESISTANCE 167

Saroj Sharma, Monalisa Tiwari and *Vishvanath Tiwari*

INTRODUCTION 167
EMERGENCE OF METALLOPROTEOMICS 168
CURRENT APPROACHES TO METALLOPROTEOMICS 170
ROLE OF METALLOPROTEINS IN MICROBIAL DRUG RESISTANCE 172
METAL INDUCIBLE MICROBIAL PROTEOME AND HOST-PATHOGEN
INTERACTION 173
TARGETING METALLOPROTEINS TO TARGET MICROBE 173
CONCLUDING REMARKS AND FUTURE ASPECTS 174
CONSENT FOR PUBLICATION 175
CONFLICT OF INTEREST 175
ACKNOWLEDGEMENTS 175
REFERENCES 175

CHAPTER 10 PROTEOMICS OF *MYCOBACTERIUM TUBERCULOSIS*: AN OVERVIEW 179
Anil Kumar Gupta, Divakar Sharma and *Amit Singh*
INTRODUCTION 180
    Advancement of Mycobacterial Research through Proteomics 182
PROTEOMICS AND PHYSIOLOGY OF *M. TUBERCULOSIS* 183
    Identification of New Enzymes 183
    New Substrates 183
    New Metabolic Pathways 183
PROTEOMICS IN DEVELOPING NOVEL BIOMARKERS 184
PROTEOMICS AND VACCINE DISCOVERY OF *MYCOBACTERIUM
TUBERCULOSIS* 185
PATHWAY OF TB VACCINE DEVELOPMENT 187
PROTEOMICS AND HOST-PATHOGEN INTERACTION 187
    Host Cells and Environments for *M. tuberculosis* 187
PROTEOMIC ANALYSIS OF *MYCOBACTERIUM TUBERCULOSIS*-INFECTED
CELLS 189
PROTEOMICS AND STRUCTURAL BIOLOGY 191
STRUCTURE-BASED DRUG DISCOVERY 192
    Relationship between Proteomics and Genomics 193
    Utility of Proteomics in Drug Discovery 193
CONCLUSION 195
CONSENT FOR PUBLICATION 195
CONFLICT OF INTEREST 195
ACKNOWLEDGEMENTS 195
REFERENCES 195

**CHAPTER 11 PROTEOME BASED INSIGHTS IN DRUG-RESISTANT *MYCOBACTERIUM
TUBERCULOSIS*** 204
Apoorva Narain, Surya Kant, Rikesh Kumar Dubey, Kanchan Srivastava and *Anand
Kumar Maurya*
INTRODUCTION 205
DRUG RESISTANT TUBERCULOSIS 206
CLINICALLY RELEVANT MTB PROTEINS 207
    Antigens of 85 Complex 207
    Mpt64 207
    ESAT-6 and CFP-10 207
PROTEOMIC ANALYSIS IN MYCOBACTERIA 208
    Sodium Dodecyl Sulphate-Poly Acrylamide Gel Electrophoresis 208
    Two Dimensional Gel Electrophoresis 209

Western Blotting	209
Enzyme-Linked Immuno Sorbent Assay	210
Immunohistochemistry	210
Mass Spectrometry	211
Isotope-coded Affinity Tag Labeling	211
Stable Isotope Labeling by/with Amino Acids in Cell Culture	212
Isobaric Tags for Relative and Absolute Quantitation	212
X-ray Crystallography	213
Nuclear Magnetic Resonance Spectroscopy	213
PROSPECTIVE DIRECTION	214
CONSENT FOR PUBLICATION	214
CONFLICT OF INTEREST	214
ACKNOWLEDGEMENTS	214
REFERENCES	214
SUBJECT INDEX	218

FOREWORD

Dr. Divakar Sharma is intending to deliver a very useful work in the form of E-book **Volume I** entitled **"Microbial Proteomics: Development in Technologies and Applications"**, in the E-book series **"Current and Future Developments in Proteomics"**, a collection of chapters on recent development on microbial proteomics related to technologies and diseases. The editor has recruited a vast array of proteomics and microbial specialists from India to write the different chapters that constitute this volume. The chapters are structured in such a way that the reader may easily find recent developments about proteomic technologies and its application to combat the microbes related to current issues, especially on antimicrobial resistance and *M.tuberculosis*. I am sure proteomics and microbiological researchers will find this compilation useful and enjoyable.

I wish editor and authors' good luck for their contributions in delivering this very timely book.

Sanjeeva Srivastava
Group Head of Proteomics Laboratory
Indian Institute of Technology, Bombay
India

PREFACE

We are honored to contribute to the information and updates on current and future developments in proteomics for young researchers around the world. We have attempted to distill the current knowledge of proteomics technologies and development with special reference to microbe's related diseases. This volume of e-book includes various contributors from India which have been affiliated to the well known Indian Institutes.

It is divided into three volumes: Volume I, Microbial Proteomics: Development in technologies and applications; Volume II, and Volume III, will be proposed in the near future. This eBook contains full-color, high quality images of the most frequent technologies and applications with a brief and comprehensive review of microbe related diseases. Each chapter includes its novelty towards the development of proteomics technologies and applications. The format is concise, well organized, and didactic, without being exhaustive. I hope and expect that this volume will facilitate in providing updated basic and specific information to young researchers.

ACKNOWLEDGEMENTS

I would like to express my gratitude to the Indian proteomics experts (Dr. Deepa Bisht, Prof. Sanjeeva Srivastava, Dr. Suman Thakur, and Prof. Bishwajit Kundu), they provided me a positive attitude to think out of the box and sustainable support with English composition and edition. I also acknowledge Dr. Srikanth Tripathy, Dr. Krishnamurthy Venkatesan, Dr. Nirmala Deo, Dr. Shripad A. Patil, and Prof. Asad U. Khan for their positive support. I would also like to express my sincere thanks to my friends and colleagues without whose contribution; it would not have been possible to complete this project. I also want to thank the staff of Bentham Science for their help and support and giving me this opportunity to publish this eBook. I also thanks the ICMR and SERB for providing funds and fellowships for my research.

Divakar Sharma
Kusuma School of Biological Sciences
Indian Institute of Technology, Delhi
India

About the Editor

Presently, Dr. Divakar Sharma is a Principal Project Scientist at the Indian Institute of Technology, Delhi, India. He has completed his M.S. degree in Biotechnology from Chatrapati Sahu Ji Maharaj University Kanpur, India and a Ph.D. degree in Biotechnology from the Jiwaji University, Gwalior, India, as well as ICMR-National JALMA Institute for Leprosy and Other Mycobacterial Diseases, Agra, India. He has completed Two-Post Doctoral Fellowships research at ICMR-National JALMA Institute for Leprosy and Other Mycobacterial Diseases, Agra, India (2015-2017) and Aligarh Muslim University, Aligarh (2017-2019). After completion of the Post Doctoral Fellowships, he joined the Indian Institute of Technology, Delhi, India as a Principal Project Scientist. He is an active Proteomics Researcher in the Antimicrobial Resistance (AMR) for the last 12 years. His research focuses on drug resistance exploration of *M.tuberculosis* and opportunistic pathogens through proteomics and bioinformatics based approaches. His contribution to this field includes more than 50 peer-reviewed articles and book chapters of international repute. He is the editorial board member for various visible journals of international repute affiliated to the well-known publishers like Bentham Science, BMC, Frontiers, and many more. He is also an expert's panelist member of several international journals. Apart from that, he is a life member of various societies in India like Proteomics Society of India, Indian Society for Mass Spectrometry, and Association of Microbiologist of India.

DEDICATION

This volume of the Ebook series is especially dedicated to Dr. Deepa Bisht, Ph.D., Scientist E, at ICMR-National JALMA Institute for Leprosy and Other Mycobacterial Diseases, Agra, India. She always inspires me to work hard, try to do my best, and think out of the box. She guided me during my doctoral and post doctoral research; and has been my mentor and treats me in a friendly manner. She is a recognized Indian scientist in the area of M. tuberculosis proteomics research.

Divakar Sharma
Kusuma School of Biological Sciences
Indian Institute of Technology, Delhi
India

List of Contributors

Abhishek Parashar	Animal Biotechnology Centre, National Dairy Research Institute, Karnal, India
Aditya Arya	Centre for bioinformatics, computational and systems biology, Pathfinder Research and Training Foundation, Greater Noida, Uttar Pradesh, India
Amit Singh	Department of Microbiology, All India Institute of Medical Sciences, Saket Nagar, Bhopal, India All India Institute of Medical Sciences, New Delhi, India
Anand Kumar Maurya	Department of Microbiology, All India Institute of Medical Sciences, Bhopal, India
Ankita Mathur	Mewar University, Gangrar, Chittorgarh, Udaipur, 312901, India
Anil Kumar Gupta	Department of Microbiology, All India Institute of Medical Sciences, Saket Nagar, Bhopal, India
Apoorva Narain	Department of Respiratory Medicine, King George Medical University, Lucknow, India
Arpana Sharma	Central University of Gujarat, Gandhinagar, India Mehsana Urban Institute of sciences, Department of Biotechnology, Ganpat University, Kherva, 384012, India
Arvind K. Verma	Animal Biotechnology Centre, National Dairy Research Institute, Karnal, India
Ashok K. Mohanty	Animal Biotechnology Centre, National Dairy Research Institute, Karnal, India
Chandrajeet Singh	Central University of Gujarat, Gandhinagar, India
Deepa Bisht	Department of Biochemistry, National JALMA Institute for Leprosy and other Mycobacterial Diseases, Tajganj, Agra 282004, India
Divakar Sharma	Kusuma School of Biological Sciences, Indian Institute of Technology, Delhi, India
Gopika Raval	Mehsana Urban Institute of sciences, Department of Biotechnology, Ganpat University, Kherva, 384012, India
Juhi Sharma	Department of Botany and Microbiology, St. Aloysius College, Sadar Cantt, Jabalpur, 482002, India
Kanchan Srivastava	Department of Respiratory Medicine, King George Medical University, Lucknow, India
Kruti Dave	Mehsana Urban Institute of sciences, Department of Biotechnology, Ganpat University, Kherva, 384012, India
M. Madhan Kumar	ICMR-National JALMA Institute for Leprosy and other Mycobacterial Diseases, Tajganj, Agra-282004, India
Monalisa Tiwari	Department of Biochemistry, Central University of Rajasthan, Ajmer-305817, India
Munna Lal Yadav	Animal Biotechnology Centre, National Dairy Research Institute, Karnal, India

vi

Nirmala Deo	Department of Biochemistry, National JALMA Institute for Leprosy and other Mycobacterial Diseases, Tajganj, Agra 282004, India
Pranav Kumar Prabhakar	Department of Medical Laboratory Sciences, Lovely Professional University, Punjab-144411, India
Preeti Rawat	Animal Biotechnology Centre, National Dairy Research Institute, Karnal, India
Rikesh Kumar Dubey	Division of Microbiology, CSIR-Central Drug Research Institute, Lucknow, India
Sandeep Sharma	Department of Medical Laboratory Sciences, Lovely Professional University, Phagwara, Punjab-144411, India
Sarika Sharma	Department of Life Science, Arni University, Kathgarh, Indora, Kangra. H.P.-176401, India
Saroj Sharma	Department of Biochemistry, Central University of Rajasthan, Ajmer-305817, India
Sudarshan Kumar	Animal Biotechnology Centre, National Dairy Research Institute, Karnal, India
Surya Kant	Department of Respiratory Medicine, King George Medical University, Lucknow, India
Vishvanath Tiwari	Department of Biochemistry, Central University of Rajasthan, Ajmer-305817, India
Vivek DharDwivedi	Centre for bioinformatics, computational and systems biolog, Pathfinder Research and Training Foundation, Greater Noida, Uttar Pradesh, India
Vivek Kumar Gupta	ICMR-National JALMA Institute for Leprosy and other Mycobacterial Diseases, Tajganj, Agra-282004, India
Yogesh K. Dhuriya	Developmental Toxicology Laboratory, Systems Toxicology and Health Risk Assessment Group,CSIR-Indian Institute of Toxicology Research (CSIR-IITR) Vishvigyan Bhawan, 31, Mahatma Gandhi Marg Lucknow–226 001, India

CHAPTER 1

Microbial Biofilm and Drug Resistance: A Proteomic Approach

Sarika Sharma[1] and **Sandeep Sharma**[2,*]

[1] *Department of Life Sciences, Arni University, Kathgarh, Indora, Kangra. H.P.-176401, India*

[2] *Department of Medical Laboratory Sciences, Lovely Professional University, Phagwara, Punjab-144411, India*

Abstract: Bacterial Biofilms are densely packed microbial communities formed by the microorganism to escape from any external threat. These Biofilms are composed of a polysaccharidic matrix, which formed a slimy layer outside the cell wall and protected the microbes from any damage (both physical and chemical). Biofilm can be developed by microorganisms on any surface, including medical devices, oral cavity and other biomaterials. These biofilms are difficult to treat and required almost 10,000 more concentrations of antibiotics compared to planktonic microorganisms. These biofilms are the main hindrance in the treatment of hospital-acquired infections. Majority of the treatment against microorganisms fails due to these biofilms in a clinical setting. Combination of different antibiotics, natural molecules and other strategies is in use to combat these biofilms. Microorganisms inside the Biofilm trigger some genes by quorum sensing and affect the expression of several protein factors. The current chapter will focus on the use of the proteomic approach for better understanding the nature and role of Biofilm in microbial pathogenesis and lower the emergence of drug resistance in these microorganisms.

Keywords: Antibiotics, Biofilm, Drug resistance, Microorganism, Pathogenesis, Polysaccharidicmatrix, Proteomics.

INTRODUCTION

Microbes are ubiquitous in nature and always busy in executing tasks. Some are building products for the benefit of humankind, and some are creating problems for humans. Biofilms are defined as organized communities of bacterial cells that are enclosed in a self-constructed polymeric matrix and adhere to any surfaces, an interface or each other [1]. Bacterial cells in biofilms are much different from their planktonic counterparts or free-floating forms. Biofilms are the extremely preferred approach of bacterial growth to survive in adverse conditions [2]. In the

---

* **Corresponding author Sandeep Sharma:** Department of Medical Laboratory Sciences, Lovely Professional University, Phagwara, Punjab-144411, India; E-mails: sandeep4380@gmail.com and sandeep.23995@lpu.co.in.

Divakar Sharma (Ed.)
All rights reserved-© 2020 Bentham Science Publishers

natural environment, biofilms are found almost everywhere, but the negative side of Biofilm lies within Biofilm associated infections in humans. Nowadays, biofilms are accountable for many human infections, and 65 to 80% of all human infections are linked with microbial biofilms [3]. These include wound infection, infection of the lower respiratory tract in cystic fibrosis, infection in kidney stones and implanted devices in orthopedic surgery. The Biofilm associated bacterial cells have a clinically significant property with their high antibiotic resistance level compared with their free-living (planktonic) counterparts [4]. The reason for increased drug resistance may be that the bacteria within the Biofilm grow more slowly than their planktonic formed or the matrix formed by sessile bacteria, preventing the entry of the drug into the cell, while transcriptomics and proteomics revealed that the gene expression in free-floating bacterial cells is much different from bacterial cells associated with Biofilm or sessile bacterial cells [5, 6]. There were reports published in the last few years that increase resistance to an antibiotic is directly related to the high level of protein expression involved in glycolysis, microbial metabolism, and synthesis of secondary metabolites [7]. Moreover, there are also some other biological pathways which are directly involved in cell to cell communications (LuxS-mediated quorum sensing, arginine metabolism, rhamnose biosynthesis). Few other proteins (pheromone and adhesion associated proteins) were found to be upregulated during the biofilm transit from planktonic stages [8]. The exact molecular mechanisms for increased drug resistance to Biofilm are not fully understood, although many studies have already been conducted for so many years. Advances in proteomics techniques during the past decade have facilitated an in-depth analysis of the possible mechanisms underpinning increased drug resistance within Biofilm. Proteomic studies are also to establish its role in better understanding the nature of bacterial biofilms. Biofilm proteomics is defined as to study the complete set of proteins of bacteria embedded in the Biofilm [9, 10]. In this chapter, various proteomics approaches related to the Biofilm, and their increased drug resistance to pathogens will be illustrated.

The Process of Biofilm Formation

Biofilm formation is an exceptionally sophisticated process in which free-floating microbial cells will be transformed into sessile form. The formation of Biofilm is a multistage process in which a series of events takes place as an adaptation of microbial cells to survive and divide in a broad range of harsh environmental conditions [11]. Biofilm formation is a multi-step process which includes attachment of free-floating microbial cells to a surface, the growth and aggregation of cells into microcolonies followed by growth into mature, structurally complex Biofilm (maturation), and the dispersal of detached bacterial cells as shown in Fig. (1) [12].

Attachment

The attachment of the bacterial cells to the surface is mediated by any one of them (like week van der Waals forces, electrostatic and hydrophilic interactions) or by a combination of all these interactions [13]. Pili and pilus-associated adhesins have also been reported to play an important role in adhesion and colonization of microbial cells to the surfaces [14, 15]. The microbial cells that initiate attachment are covered by only small amounts of exopolymeric material. While following the initial attachment to a surface, the bacteria have to maintain contact with the surface. Thus, the stage of attachment is followed by a phase during which the synthesis of bacterial exopolysaccharides (EPS) provides steady attachment by forming organic bridges between the cells and the surface [16].

Micro-Colony Formation

After the bacteria adhered to the surface, the bacterial cell multiplies within the embedded exopolysaccharide matrix, which lastly results in micro-colony formation. After the micro-colony formation stage of Biofilm, there is an upregulation of certain biofilm genes. It is observed that bacterial attachment can trigger the formation of an extracellular matrix [17]. Matrix formation is followed by water-filled channels formation for the transport of nutrients within the Biofilm. Researcher has proposed that these water channels are like a circulatory system, distributing different nutrients to and removing waste materials from the communities in the micro-colonies of the Biofilm [3, 18].

Detachment

The growth of the bacterial Biofilm is eventually restricted by the availability of nutrients accumulation of toxic by-products and other factors, including pH, oxygen perfusion, carbon source availability and osmolarity [19]. At some point, dispersing of biofilm cells occur either by the detachment of newly formed cells from growing cells or dispersion of biofilm aggregates. The cells may detach, and together with the progeny of other biofilm cells, may colonize other surfaces [20]. Enzymes such as polysaccharide lyases that degrade the extracellular polysaccharide matrix have been reported to play a role in biofilm dissolution in several organisms [21]. The cells which are dispersed from biofilm as a result of growth may return quickly to their normal planktonic phenotype.

Mechanism of Biofilm Resistance against Antimicrobials

Resistance to drugs and antibiotics is attained by microbes permanently or temporarily during the survival strategy of microbes in adverse circumstances [22]. Antibiotic resistance is nowadays a serious threat for humankind as the present regime of antimicrobials is not working against existing resistant microbes [23]. It has been reported that the sessile form of microbes is harder to treat than the planktonic ones. There are many mechanisms shown by sessile microbes against antimicrobials [24].

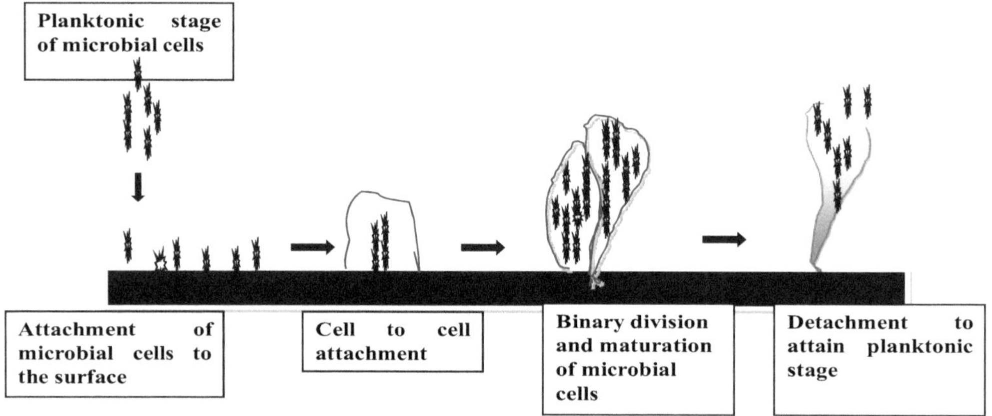

Fig. (1). Steps in biofilm formation.

Limited Penetration of Drugs Due to Glycocalyx

Microbes present in biofilms have increased resistance towards many drugs and biocides, and researchers have shown that the presence of glycocalyx partially inhibits the entry of drugs by forming a diffusion barrier to the antimicrobial agents [25]. Glycocalyx is an integral component of the biofilms, and its thickness varied from 0.2 to 1.0μm. The time of the entry of the drug into cells takes more time as the distance to the cell increases due to the presence of glycocalyx, by molecular sieving (size exclusion), and by the viscosity of glycocalyx. Nonetheless, limited penetration of drugs into the bacterial cells *via* glycocalyx is one of the most probable resistance mechanisms associated with biofilms [22, 26].

Enzyme-Mediated Resistance

Resistance to antimicrobials or drugs can be attributed to the presence of enzymes that transform bactericidal compounds into the nontoxic form [27]. There are some species of bacteria known for the degradation of toxic compounds such as aromatic, phenolic and other heavy metals [28]. The process of detoxification is done by enzymatic reduction, which converts the cations to the metal, whereas some plasmids harbor heavy metal resistance genes [29]. The presence of heavy metals in the bacterial cells induces the enzymes and results in drug resistance. Some heavy metals induce resistance to a broader spectrum of antibiotics [30].

State of Metabolism and Rate of Growth in Biofilms

It was found that the growth rate of sessile bacteria is much slower than their planktonic counter parts [4]. The metabolic activities of cells were enhanced by nutrients and oxygen in the biofilms, which sustains bacterial growth and metabolism [31, 32]. Biofilm increases the level of resistance against antibiotics under limiting the concentration of oxygen and nutrients by expressing some genes under anaerobic conditions [33]. While due to poor diffusion of nutrients, the metabolic activities of bacterial cells slow down inside Biofilm, leading to slowly growing cells [34]. When biofilms are treated with the inhibitory concentration of any bactericidal agents, the cells within biofilms lead to phenotypic adaption and show resistance. The factors such as heat or starvation stress enhance tolerance of bacterial biofilms against antimicrobial agents [35, 36].

Genetic Adaptations

The formation of Biofilm is an adaption strategy of bacteria under harsh conditions so, during biofilm formation, bacterial cells undergo many phenotypic changes [37]. These phenotypic changes are accompanied by genotype changes that make the sessile bacterial cell more robust to treat with antimicrobials. The multiple antibiotic resistance (*mar*) operons are known regulator which controls the expression of *mar* genes in *E. coli.* encodes for multi-drug resistant phenotype and shows resistance to various antibiotics, organic solvents and disinfectants [38]. Variety of genes have been reported in *E. coli* responsible for drug resistance like superoxide dismutases, catalysts, alkyl hydroperoxide reductases and glutathione reductases, and enzymes involved in the DNA repair mechanism of the bacterial cell [39]. In addition, various regulatory genes such as *oxyR* and *sox are* also reported to activate when cells are exposed to stress conditions such as oxidizing agents and thus provide resistance to many drugs [40].

Efflux Pumps

Efflux pumps are the key component of the low level of drug resistance in bacteria [41]. Efflux pumps favor bacterial vitality in adverse conditions, including antimicrobial agents [42]. Efflux pumps attributed both intrinsic and acquired resistance of bacterial cells against antibacterial agents that are structurally related or different. There are five different classes of bacterial efflux pumps identified that are accountable for drug resistance [43]: (1) the major facilitator superfamily (MF), (2) the resistance-nodulation-division family (RND), (3) the small multidrug resistance family (SMR), (4) the ATP-binding cassette family (ABC) and the (5) multidrug and toxic compound extrusion family (MATE) [44]. The ABC family system hydrolyses ATP to work as antimicrobial agent efflux, while the MF family, MATE family, and the RND family served as secondary transporters, catalyzing drug ion antiproton (H^+ or Na^+) [45]. RND family transporters mutate the drug target or drug in the bacterial cell [46]. The lower concentration of many antibiotics, such as chloramphenicol and tetracycline and xenobiotics such as salicylate and chlorinated phenols, activates the expression of multi-drug resistance operons and efflux pumps in the bacterial cells entrapped in biofilms [43 - 47].

Quorum Sensing

Quorum sensing (QS) is a process of cell-to-cell communication *via* specific signals which are given from one cell to another cell in bacteria for their survival under adverse conditions [48]. The QS signals in the bacteria mainly consist of acyl-homoserine lactone (AHLs), autoinducing peptide (AIPs), and autoinducer-2 (AI-2), which play a major role in the pathogenesis mechanism of bacteria [49]. The activation of QS signaling aids biofilm formation and leads to antimicrobial resistance [50]. QS signals participate in the synthesis of various virulence factors such as lectin, exotoxin A, pyocyanin, and elastase in the *Pseudomonas aeruginosa* during its growth and infection [51]. These virulence factors are regulated by QS and help bacterial cells to invade the host immune and obtain nutrition from the host. The Gram-negative and Gram-positive bacterial cells have different QS signals for cell-to-cell interactions. The Gram-negative bacteria mainly produce the AHL signaling molecules [52], and Gram-positive bacterial cells produce AIP signaling molecules for the cell to cell communication [53].

TARGET PROTEINS TO STUDY BIOFILM

Proteins Involved in Cellular Metabolism

There are some basic cellular processes occurred in the bacterial cell, which are crucial in biofilm formation, such as glycolysis, Kreb's cycle, protein synthesis and its degradation, fatty acid synthesis and its degradation, *etc* [54]. Motility proteins have a key role in biofilm formation, as reported by Moorthy and Watnick [55]. Biofilms are initiated by attachment of bacteria to the surface, and the process of attachment requires proteins of flagella and pili for adhesion, as reported in *Escherichia coli, Salmonella enterica, Vibrio cholerae*, and *Pseudomonas aeruginosa* [56]. It was found that the proteins responsible for the synthesis of flagella are downregulated and slow down the movement of bacterial cells and lead to the development of Biofilm [57].

Transporter Proteins

Many studies on planktonic and sessile bacterial cells reported that the proteins involved in the transport of molecules in the microbial cells have a major role in biofilm formation [58]. Most of the transporter proteins are located in the membrane and served as primary transporters and secondary transporters and worked as channels and carriers [59]. Expression of various transporters (ABC transporters) is altered accordingly to the exterior surroundings, as reported in *Bacillus subtilis, Staphylococcus aureus, T. maritime, E. coli, Streptococcus mutans* and *P. aeruginosa* [60 - 64]. Differential expression of these transport proteins under diverse conditions is essential for the growth of bacteria as well as the formation and stability of Biofilm [65].

Stress-responsive Proteins

It is now well known that biofilm formation is done by microbial cells to survive under harsh conditions. This feature of bacteria requires the expression of some specific proteins that aid bacterial cells to adapt themselves successfully in adverse conditions. In the process of biofilm formation, researchers reported that there is 10 to 20-fold up-regulation of oxidative stress-related proteins [66], which contributed resistance and protection from oxidative stress. The two more oxidative stresses related proteins, alkyl hydroperoxide reductase and thioredoxin, were reported to be over-expressed in the Biofilm formed by *Helicobacter pylori* [67]. Two chaperone proteins GroES and GroEL and ß-subunit of RNA polymerase were also up-regulated under serum starvation [67].

Proteomic Techniques to Study Biofilm

In the last two decades, drug resistance has emerged as a serious threat to the health of the human race and to figure out the mechanism behind antibiotic resistance is imperative to boost the efficacy of current antibiotics and to explore novel targets [68]. The study of protein profiles by proteomic analysis has become a crucial tool for the study of underlying mechanisms related to bacterial resistance and virulence. In proteomics, analysis of the total protein content of cells is carried out. The proteomic response is peculiar for every antibiotic, but proteins involved in energy and nitrogen metabolism, protein and nucleic acid synthesis, glucan biosynthesis, and stress response are often effected [69].

The proteomic techniques are regularly being advanced and an immense variety of methods and applications are available [8 - 70]. Bacterial proteome analyses commenced with traditional electrophoretic techniques like sodium dodecyl sulphate polyacrylamide electrophoresis (SDS-PAGE). SDS-PAGE is an analytical technique to separate proteins based on their molecular weight. The ionic detergent SDS denatures the proteins and makes them uniformly negatively charged [71]. Thus, when the current is applied, all SDS-bound proteins in a sample will migrate towards the cathode and proteins with lower mass move more quickly through the gel than those with larger mass because of the sieving effect of the gel. After SDS-PAGE, most quantitative proteomic studies were performed by 2D-PAGE, 2-DE, which is a form of gel electrophoresis commonly used to analyze mixtures of proteins [72]. In two-dimensional gel electrophoresis, proteins are first separated based on their pI (isoelectric point) in isoelectric focusing and then further separated by molecular weight through SDS-PAGE; thus, the sample proteins are distributed across the two-dimensional gel profile. This technique broadens the number of proteins that can be studied and produce more efficient data and detailed information for proteomics analysis, particularly after the incorporation of immobilized pH gradients and of the difference fluorescent labeling method of electrophoresis (DIGE). The gel-free methods become more prominent, and metabolic labeling was included in quantitative analysis.

Stable Isotope Labeling by Amino Acids in Cell Culture (SILAC)

SILAC is a metabolic labeling technique for mass spectrometric based quantitative proteomics. Cells are differentially labeled by growing them in media supplied with either a light or heavy stable isotope of a nutrient, generally a nitrogen source or an amino acid [73]. Metabolic incorporation of amino acids into the proteins results in the mass shift of the corresponding peptides; this mass shift is detected by the mass spectrometer. Typical SILAC labeling uses lysine and arginine, which in combination with trypsin digestion, results in labeling every peptide in the mixture; the so-called stable isotope labeling by amino acids in cell culture (SILAC) strategy now prevails [74, 75].

Isotope-coded Affinity Tags (ICAT)

As a substitute for metabolic labeling, differential analysis can be carried out by chemical labeling. Isotope-coded affinity tags (ICAT) consist of three functional elements: a specific chemical reactive group that binds to sulfhydryl groups of cysteinyl residues, an isotopically coded linker with light or heavy isotopes, and a biotin tag for affinity purification [76]. The cysteine residue of extracted labeled with either a light or heavy (deuterium containing). Afterwards, light and heavy labeled samples are pooled and proteolytically cleaved. Later, the complexity of the sample is reduced prior to MS analysis through the purification of tagged cysteine-containing peptides by affinity chromatography using biotin-avidin affinity columns. Peptide pairs with 8 Da mass-shifts are detected in MS scans, and their ion intensities are compared for relative quantitation [77]. ICAT labelling takes place at the protein level allowing samples to be pooled prior to protease treatment. On the other hand, cysteine is not found very much in proteins, and approximately one in seven proteins do not contain this amino acid, greatly reducing the efficacy of the process [78].

Isobaric Tags for Relative and Absolute Quantification (iTRAQ)

Today, the multiple isobaric tags are most preferred among chemical isotope labeling methods [79]. iTRAQ tags are isobarics and mainly designed for the labeling of peptides rather than proteins. The overall molecule mass is kept constant at 145 Da for iTRAQ-4plex and 304 Da for -8plex. The iTRAQ-4plex molecule contains a reporter group (based on N-methylpiperazine), a mass balance group (carbonyl), and a peptide-reactive group (NHS ester). The iTRAQ reagents label N-terminal peptide and ε-amino groups of lysine side chains and allow comparison of up to eight samples in the same experiment. The iTRAQ labelled peptides are further purified by reverse-phase liquid chromatography and then analyzed by tandem MS [80]. Each tag releases a distinct mass reporter ion upon peptide fragmentation, the ratio of which determines the relative abundances of the peptides iTRAQ labeled peptides appear as a single unresolved precursor at the same m/z in the MS spectrum. Upon peptide fragmentation, the iTRAQ labeled fragments produce reporter ions in a "silent region," usually unpopulated, at low m/z range (*e.g.*, 114–121). Measurements of the reporter ion intensities enable the relative quantification of the peptide in each sample [78].

Tandem Mass Tags (TMT)

A tandem mass tag is an MS/MS-based quantitative method using isotopomer labels in iTRAQ [81]. Both techniques share several common features. (i) These methods use reagent N-hydroxy succinimide (NHS), which does specific tagging of primary amino groups. (ii) They were designed to allow multiplexing of several samples by chemical derivatization with different forms of the same isobaric tag that appear as a single peak in MS analysis. (iii) The release of "daughter ions" in MS/MS analysis weighs between 126 and 131 Da that can be used for relative quantification. The cysteine reactive TMT (cysTMT) reagents enable selective labeling and relative quantitation of cysteine-containing peptides from up to six biological samples [82, 83].

Isotope-coded Protein Label (ICPL)

In the technique of Isotope-coded protein label (ICPL), free amino groups present in proteins are isotopically labeled [84]. Two protein mixtures are reduced and alkylated for easy access to free amino groups that are subsequently labelled with the isotopes of deuterium-free (light) or 4 deuteriums containing (heavy). Light-

and heavy-labeled samples are then mixed, fractionated, and digested before high throughput MS analysis [85]. The peptides of identical sequences derived from the two deferentially labeled protein samples of different mass, they appear as doublets in the acquired MS spectra. From the ratios of the ion intensities of these peptide pairs, the relative abundance of their parent proteins in the original samples can be determined [86].

Bioorthogonal Noncanonical Amino Acid Tagging (BONCAT)

It is a leading proteome labeling technique that is highly sensitive and based on the *in vivo* incorporation of synthetic amino acids that exploit the substrate promiscuity of artificial amino acids such as azidohomoalanine (AHA) or homopropargylglycine (HPG). Both molecules compete with methionine and are incorporated into newly synthesized protein [87]. When the cell is anabolically active, AHA and HPG are detected *via* azide-alkyne click chemistry. This method depends on pulse labeling and helps in identifying the *de novo* synthesized proteome of microbial cells, which are translationally active [88].

Future Prospectus

Proteomics, the promising new "omics," has emerged as an important integral tool to genomics, providing novel information and greater vision into the proteome of bacterial biofilms. Advancement in proteomics techniques like MS and metabolic labeling, together with bioinformatics tools, can enhance our knowledge about proteins that can be targeted to combat with increased resistance of microbes present in biofilms. More studies are required to generate information about the regulation of proteins involved in biofilm formation. Metaproteomics and proteogenomics approaches can improve our knowledge in understanding proteomes of sessile microbes and aid in the identification of target proteins accountable for biofilm formation and resistance towards antibiotics. In the future, proteomic technology will become an asset for researchers and scientists for diagnosing infections, discovery of novel antibiotics and innovative therapies against drug-resistant microbes.

CONCLUSION

Nowadays, the drug resistance of microbes and the study of microbial Biofilm are one of the highly crucial affairs for the scientific community. The rapidly developing field of genomics transcriptomics and proteomics elucidated the complexity of biofilm regulation. The combination of various proteomics techniques provides a detailed understanding of the proteins involved at every step of biofilm formation, starting from the attachment of microbial cells to the surface, their development, and the detachment from the surface. In due course, the proteomics study on biofilm formation will lead to improved strategies for inhibition of biofilm formation and decipher the protein targets for the treatment of various infections caused by biofilms.

CONSENT FOR PUBLICATION

Not Applicable.

CONFLICT OF INTEREST

The authors declare no conflict of interest, financial or otherwise.

ACKNOWLEDGEMENTS

Declared none.

REFERENCES

[1]   Costerton JW, Lewandowski Z, Caldwell DE, Korber DR, Lappin-Scott HM. Microbial biofilms. Annu Rev Microbiol 1995; 49: 711-45.
[http://dx.doi.org/10.1146/annurev.mi.49.100195.003431] [PMID: 8561477]

[2]   Kostakioti M, Hadjifrangiskou M, Hultgren SJ. Bacterial biofilms: development, dispersal, and therapeutic strategies in the dawn of the postantibiotic era. Cold Spring Harb Perspect Med 2013; 3(4)a010306
[http://dx.doi.org/10.1101/cshperspect.a010306] [PMID: 23545571]

[3]   Jamal M, Ahmad W, Andleeb S, *et al.* Bacterial biofilm and associated infections. J Chin Med Assoc 2018; 81(1): 7-11.
[http://dx.doi.org/10.1016/j.jcma.2017.07.012] [PMID: 29042186]

[4]   Hall CW, Mah TF. Molecular mechanisms of biofilm-based antibiotic resistance and tolerance in pathogenic bacteria. FEMS Microbiol Rev 2017; 41(3): 276-301.
[http://dx.doi.org/10.1093/femsre/fux010] [PMID: 28369412]

[5]   Vilain S, Cosette P, Zimmerlin I, Dupont JP, Junter GA, Jouenne T. Biofilm proteome: homogeneity or versatility? J Proteome Res 2004; 3(1): 132-6.
[http://dx.doi.org/10.1021/pr034044t] [PMID: 14998174]

[6]   Sharma D, Misba L, Khan AU. Antibiotics *versus* biofilm: an emerging battleground in microbial communities. Antimicrob Resist Infect Control 2019; 8(1): 76.
[http://dx.doi.org/10.1186/s13756-019-0533-3] [PMID: 31131107]

[7]   Li W, Yao Z, Sun L, *et al.* Proteomics analysis reveals a potential antibiotic cocktail therapy strategy for Aeromonas hydrophila infection in Biofilm. J Proteome Res 2016; 15(6): 1810-20.
[http://dx.doi.org/10.1021/acs.jproteome.5b01127] [PMID: 27110028]

[8]   Suryaletha K, Narendrakumar L, John J, Radhakrishnan MP, George S, Thomas S. Decoding the proteomic changes involved in the biofilm formation of *Enterococcus faecalis* SK460 to elucidate potential biofilm determinants. BMC Microbiol 2019; 19(1): 146.
[http://dx.doi.org/10.1186/s12866-019-1527-2] [PMID: 31253082]

[9]   Khemiri A, Jouenne T, Cosette P. Proteomics dedicated to biofilmology: What have we learned from a decade of research? Med Microbiol Immunol (Berl) 2016; 205(1): 1-19.
[http://dx.doi.org/10.1007/s00430-015-0423-0] [PMID: 26068406]

[10]   Sharma D, Garg A, Kumar M, Rashid F, Khan AU. Down-Regulation of Flagellar, Fimbriae, and Pili

Proteins in Carbapenem-Resistant *Klebsiella pneumoniae* (NDM-4) Clinical Isolates: A Novel Linkage to Drug Resistance. Front Microbiol 2019; 10: 2865.
[http://dx.doi.org/10.3389/fmicb.2019.02865] [PMID: 31921045]

[11]   Yin W, Wang Y, Liu L, He J. Biofilms: The Microbial "Protective Clothing" in Extreme Environments. Int J Mol Sci 2019; 20(14): 3423.
[http://dx.doi.org/10.3390/ijms20143423] [PMID: 31336824]

[12]   Okada M, Sato I, Cho SJ, *et al.* Structure of the Bacillus subtilis quorum-sensing peptide pheromone ComX. Nat Chem Biol 2005; 1(1): 23-4.
[http://dx.doi.org/10.1038/nchembio709] [PMID: 16407988]

[13]   Oh JK, Yegin Y, Yang F, *et al.* The influence of surface chemistry on the kinetics and thermodynamics of bacterial adhesion. Sci Rep 2018; 8(1): 17247.
[http://dx.doi.org/10.1038/s41598-018-35343-1] [PMID: 30467352]

[14]   Bullitt E, Makowski L. Structural polymorphism of bacterial adhesion pili. Nature 1995; 373(6510): 164-7.
[http://dx.doi.org/10.1038/373164a0] [PMID: 7816100]

[15]   Chahales P, Thanassi DG. Structure, function, and assembly of adhesive organelles by uropathogenic bacteria. Urinary Tract Infections: Molecular Pathogenesis and Clinical Management 2017, Feb; 15: 277-329.
[http://dx.doi.org/10.1128/9781555817404.ch14]

[16]   Wolska KI, Grudniak AM, Rudnicka Z, Markowska K. Genetic control of bacterial biofilms. J Appl Genet 2016; 57(2): 225-38.
[http://dx.doi.org/10.1007/s13353-015-0309-2] [PMID: 26294280]

[17]   Steinberg N, Kolodkin-Gal I. The matrix reloaded: how sensing the extracellular matrix synchronizes bacterial communities
[http://dx.doi.org/10.1128/JB.02516-14]

[18]   Chandki R, Banthia P, Banthia R. Biofilms: A microbial home. J Indian Soc Periodontol 2011; 15(2): 111-4.
[http://dx.doi.org/10.4103/0972-124X.84377] [PMID: 21976832]

[19]   Sauer K, Camper AK. Characterization of phenotypic changes in *Pseudomonas putida* in response to surface-associated growth. J Bacteriol 2001; 183(22): 6579-89.
[http://dx.doi.org/10.1128/JB.183.22.6579-6589.2001] [PMID: 11673428]

[20]   Berne C, Kysela DT, Brun YV. A bacterial extracellular DNA inhibits settling of motile progeny cells within a biofilm. Mol Microbiol 2010; 77(4): 815-29.
[http://dx.doi.org/10.1111/j.1365-2958.2010.07267.x] [PMID: 20598083]

[21]   Kaplan JB. Biofilm dispersal: mechanisms, clinical implications, and potential therapeutic uses. J Dent Res 2010; 89(3): 205-18.
[http://dx.doi.org/10.1177/0022034509359403] [PMID: 20139339]

[22]   Singh S, Singh SK, Chowdhury I, Singh R. Understanding the mechanism of bacterial biofilms resistance to antimicrobial agents. The Open Microbiology Journals 2017; 11: 53.
[http://dx.doi.org/10.2174/1874285801711010053]

[23]   Li B, Webster TJ. Bacteria antibiotic resistance: New challenges and opportunities for implant-associated orthopedic infections. J Orthop Res 2018; 36(1): 22-32.
[PMID: 28722231]

[24]   Yasir M, Willcox MDP, Dutta D. Action of antimicrobial peptides against bacterial biofilms. Materials (Basel) 2018; 11(12): 2468.
[http://dx.doi.org/10.3390/ma11122468] [PMID: 30563067]

[25]   Brown ML, Aldrich HC, Gauthier JJ. Relationship between glycocalyx and povidone-iodine resistance in *Pseudomonas aeruginosa* (ATCC 27853) biofilms. Appl Environ Microbiol 1995; 61(1): 187-93.

[http://dx.doi.org/10.1128/AEM.61.1.187-193.1995] [PMID: 7887601]

[26]   Childs SG. Biofilm: the pathogenesis of slime glycocalyx. Orthop Nurs 2008; 27(6): 361-9.
        [http://dx.doi.org/10.1097/01.NOR.0000342424.51960.ce] [PMID: 19057364]

[27]   Egorov AM, Ulyashova MM, Rubtsova MY. Bacterial enzymes and antibiotic resistance. Acta
        Naturae 2018; 10(4): 33-48.
        [http://dx.doi.org/10.32607/20758251-2018-10-4-33-48] [PMID: 30713760]

[28]   Ojuederie OB, Babalola OO. Microbial and plant-assisted bioremediation of heavy metal polluted
        environments: a review. Int J Environ Res Public Health 2017; 14(12): 1504.
        [http://dx.doi.org/10.3390/ijerph14121504] [PMID: 29207531]

[29]   Oves M, Saghir Khan M, Huda Qari A, Nadeen Felemban M, Almeelbi T. Heavy metals: biological
        importance and detoxification strategies. J Bioremediat Biodegrad 2016; 7(2): 1-5.

[30]   Jan AT, Azam M, Siddiqui K, Ali A, Choi I, Haq QM. Heavy metals and human health: mechanistic
        insight into toxicity and counter defense system of antioxidants. Int J Mol Sci 2015; 16(12): 29592-
        630.
        [http://dx.doi.org/10.3390/ijms161226183] [PMID: 26690422]

[31]   Patel I, Patel V, Thakkar A, Kothari V. Microbial biofilms: microbes in social mode. Ir J Agric Food
        Res 2014; 3(2)

[32]   Żur J, Wojcieszyńska D, Guzik U. Metabolic responses of bacterial cells to immobilization. Molecules
        2016; 21(7): 958.
        [http://dx.doi.org/10.3390/molecules21070958] [PMID: 27455220]

[33]   Bester E, Kroukamp O, Wolfaardt GM, Boonzaaier L, Liss SN. Metabolic differentiation in biofilms
        as indicated by carbon dioxide production rates. Appl Environ Microbiol 2010; 76(4): 1189-97.
        [http://dx.doi.org/10.1128/AEM.01719-09] [PMID: 20023078]

[34]   Stewart PS. Mechanisms of antibiotic resistance in bacterial biofilms. Int J Med Microbiol 2002;
        292(2): 107-13.
        [http://dx.doi.org/10.1078/1438-4221-00196] [PMID: 12195733]

[35]   Stewart PS. Antimicrobial tolerance in biofilms. Microbial biofilms 2015 Oct; 7: 269-85.
        [http://dx.doi.org/10.1128/9781555817466.ch13]

[36]   Fux CA, Costerton JW, Stewart PS, Stoodley P. Survival strategies of infectious biofilms. Trends
        Microbiol 2005; 13(1): 34-40.
        [http://dx.doi.org/10.1016/j.tim.2004.11.010] [PMID: 15639630]

[37]   Beitelshees M, Hill A, Jones CH, Pfeifer BA. Phenotypic Variation during Biofilm Formation:
        Implications for Anti-Biofilm Therapeutic Design. Materials (Basel) 2018; 11(7): 1086.
        [http://dx.doi.org/10.3390/ma11071086] [PMID: 29949876]

[38]   Maira-Litrán T, Allison DG, Gilbert P. Expression of the multiple antibiotic resistance operon (mar)
        during growth of *Escherichia coli* as a biofilm. J Appl Microbiol 2000; 88(2): 243-7.
        [http://dx.doi.org/10.1046/j.1365-2672.2000.00963.x] [PMID: 10735992]

[39]   Farr SB, Kogoma T. Oxidative stress responses in *Escherichia coli* and *Salmonella typhimurium*.
        Microbiol Rev 1991; 55(4): 561-85.
        [http://dx.doi.org/10.1128/MMBR.55.4.561-585.1991] [PMID: 1779927]

[40]   Gu M, Imlay JA. The SoxRS response of *Escherichia coli* is directly activated by redox-cycling drugs
        rather than by superoxide. Mol Microbiol 2011; 79(5): 1136-50.
        [http://dx.doi.org/10.1111/j.1365-2958.2010.07520.x] [PMID: 21226770]

[41]   Alcalde-Rico M, Hernando-Amado S, Blanco P, Martínez JL. Multidrug efflux pumps at the crossroad
        between antibiotic resistance and bacterial virulence. Front Microbiol 2016; 7: 1483.
        [http://dx.doi.org/10.3389/fmicb.2016.01483] [PMID: 27708632]

[42]   Kvist M, Hancock V, Klemm P. Inactivation of efflux pumps abolishes bacterial biofilm formation.

Appl Environ Microbiol 2008; 74(23): 7376-82.
[http://dx.doi.org/10.1128/AEM.01310-08] [PMID: 18836028]

[43]   Blanco P, Hernando-Amado S, Reales-Calderon JA, *et al.* Bacterial multidrug efflux pumps: much more than antibiotic resistance determinants. Microorganisms 2016; 4(1): 14.
[http://dx.doi.org/10.3390/microorganisms4010014] [PMID: 27681908]

[44]   Kumar A, Schweizer HP. Bacterial resistance to antibiotics: active efflux and reduced uptake. Adv Drug Deliv Rev 2005; 57(10): 1486-513.
[http://dx.doi.org/10.1016/j.addr.2005.04.004] [PMID: 15939505]

[45]   Poole K. Efflux-mediated antimicrobial resistance. J Antimicrob Chemother 2005; 56(1): 20-51.
[http://dx.doi.org/10.1093/jac/dki171] [PMID: 15914491]

[46]   Routh MD, Zalucki Y, Su CC, *et al.* Efflux pumps of the resistance-nodulation-division family: a perspective of their structure, function, and regulation in gram-negative bacteria. Adv Enzymol Relat Areas Mol Biol 2011; 77: 109-46.
[http://dx.doi.org/10.1002/9780470920541.ch3] [PMID: 21692368]

[47]   Soto SM. Role of efflux pumps in the antibiotic resistance of bacteria embedded in a biofilm. Virulence 2013; 4(3): 223-9.
[http://dx.doi.org/10.4161/viru.23724] [PMID: 23380871]

[48]   Rutherford ST, Bassler BL. Bacterial quorum sensing: its role in virulence and possibilities for its control. Cold Spring Harb Perspect Med 2012; 2(11)a012427
[http://dx.doi.org/10.1101/cshperspect.a012427] [PMID: 23125205]

[49]   Jiang Q, Chen J, Yang C, Yin Y, Yao K. Quorum sensing: a prospective therapeutic target for bacterial diseases. BioMed Research International 2019; 2019
[http://dx.doi.org/10.1155/2019/2015978]

[50]   Subhadra B, Kim DH, Woo K, Surendran S, Choi CH. Control of biofilm formation in healthcare: Recent advances exploiting quorum-sensing interference strategies and multidrug efflux pump inhibitors. Materials (Basel) 2018; 11(9): 1676.
[http://dx.doi.org/10.3390/ma11091676] [PMID: 30201944]

[51]   Lee J, Zhang L. The hierarchy quorum sensing network in *Pseudomonas aeruginosa.* Protein Cell 2015; 6(1): 26-41.
[http://dx.doi.org/10.1007/s13238-014-0100-x] [PMID: 25249263]

[52]   Talagrand-Reboul E, Jumas-Bilak E, Lamy B. The social life of Aeromonas through biofilm and quorum sensing systems. Front Microbiol 2017; 8: 37.
[http://dx.doi.org/10.3389/fmicb.2017.00037] [PMID: 28163702]

[53]   Verbeke F, De Craemer S, Debunne N, *et al.* Peptides as quorum sensing molecules: measurement techniques and obtained levels *in vitro* and *in vivo.* Front Neurosci 2017; 11: 183.
[http://dx.doi.org/10.3389/fnins.2017.00183] [PMID: 28446863]

[54]   Pisithkul T, Schroeder JW, Trujillo EA, *et al.* Metabolic remodeling during biofilm development of Bacillus subtilis. MBio 2019; 10(3): e00623-19.
[http://dx.doi.org/10.1128/mBio.00623-19] [PMID: 31113899]

[55]   Moorthy S, Watnick PI. Genetic evidence that the *Vibrio cholerae* monolayer is a distinct stage in biofilm development. Mol Microbiol 2004; 52(2): 573-87.
[http://dx.doi.org/10.1111/j.1365-2958.2004.04000.x] [PMID: 15066042]

[56]   Petrova OE, Sauer K. Sticky situations: key components that control bacterial surface attachment. J Bacteriol 2012; 194(10): 2413-25.
[http://dx.doi.org/10.1128/JB.00003-12] [PMID: 22389478]

[57]   Guttenplan SB, Kearns DB. Regulation of flagellar motility during biofilm formation. FEMS Microbiol Rev 2013; 37(6): 849-71.
[http://dx.doi.org/10.1111/1574-6976.12018] [PMID: 23480406]

[58]   Beloin C, Roux A, Ghigo JM. *Escherichia coli* biofilms. Bacterial Biofilms. Berlin, Heidelberg: Springer 2008; pp. 249-89.

[59]   Gelfand MS, Rodionov DA. Comparative genomics and functional annotation of bacterial transporters. Phys Life Rev 2008; 5(1): 22-49.
[http://dx.doi.org/10.1016/j.plrev.2007.10.003]

[60]   Sauer K, Camper AK, Ehrlich GD, Costerton JW, Davies DG. *Pseudomonas aeruginosa* displays multiple phenotypes during development as a biofilm. J Bacteriol 2002; 184(4): 1140-54.
[http://dx.doi.org/10.1128/jb.184.4.1140-1154.2002] [PMID: 11807075]

[61]   Schembri MA, Kjaergaard K, Klemm P. Global gene expression in *Escherichia coli* biofilms. Mol Microbiol 2003; 48(1): 253-67.
[http://dx.doi.org/10.1046/j.1365-2958.2003.03432.x] [PMID: 12657059]

[62]   Pysz MA, Conners SB, Montero CI, *et al.* Transcriptional analysis of biofilm formation processes in the anaerobic, hyperthermophilic bacterium *Thermotoga maritima.* Appl Environ Microbiol 2004; 70(10): 6098-112.
[http://dx.doi.org/10.1128/AEM.70.10.6098-6112.2004] [PMID: 15466556]

[63]   Shemesh M, Tam A, Steinberg D. Differential gene expression profiling of Streptococcus mutans cultured under biofilm and planktonic conditions. Microbiology (Reading) 2007; 153(Pt 5): 1307-17.
[http://dx.doi.org/10.1099/mic.0.2006/002030-0] [PMID: 17464045]

[64]   Shemesh M, Tam A, Steinberg D. Expression of biofilm-associated genes of Streptococcus mutans in response to glucose and sucrose. J Med Microbiol 2007; 56(Pt 11): 1528-35.
[http://dx.doi.org/10.1099/jmm.0.47146-0] [PMID: 17965356]

[65]   Alav I, Sutton JM, Rahman KM. Role of bacterial efflux pumps in biofilm formation. J Antimicrob Chemother 2018; 73(8): 2003-20.
[http://dx.doi.org/10.1093/jac/dky042] [PMID: 29506149]

[66]   Pham TK, Roy S, Noirel J, Douglas I, Wright PC, Stafford GP. A quantitative proteomic analysis of biofilm adaptation by the periodontal pathogen *Tannerella forsythia.* Proteomics 2010; 10(17): 3130-41.
[http://dx.doi.org/10.1002/pmic.200900448] [PMID: 20806225]

[67]   Shao C, Sun Y, Wang N, *et al.* Changes of proteome components of Helicobacter pylori biofilms induced by serum starvation. Mol Med Rep 2013; 8(6): 1761-6.
[http://dx.doi.org/10.3892/mmr.2013.1712] [PMID: 24100704]

[68]   Patel R. Biofilms and antimicrobial resistance. Clin Orthop Relat Res 2005; (437): 41-7.
[http://dx.doi.org/10.1097/01.blo.0000175714.68624.74] [PMID: 16056024]

[69]   Park AJ, Krieger JR, Khursigara CM. Survival proteomes: the emerging proteotype of antimicrobial resistance. FEMS Microbiol Rev 2016; 40(3): 323-42.
[http://dx.doi.org/10.1093/femsre/fuv051] [PMID: 26790948]

[70]   Van Oudenhove L, Devreese B. A review on recent developments in mass spectrometry instrumentation and quantitative tools advancing bacterial proteomics. Appl Microbiol Biotechnol 2013; 97(11): 4749-62.
[http://dx.doi.org/10.1007/s00253-013-4897-7] [PMID: 23624659]

[71]   Gallagher SR. SDS-polyacrylamide gel electrophoresis (SDS-PAGE). Curr Protoc Essent Lab Tech 2012; 6(1): 7-3.

[72]   Magdeldin S, Enany S, Yoshida Y, *et al.* Basics and recent advances of two dimensional-polyacrylamide gel electrophoresis. Clin Proteomics 2014; 11(1): 16.
[http://dx.doi.org/10.1186/1559-0275-11-16] [PMID: 24735559]

[73]   Gouw JW, Krijgsveld J, Heck AJ. Quantitative proteomics by metabolic labeling of model organisms. Mol Cell Proteomics 2010; 9(1): 11-24.

[http://dx.doi.org/10.1074/mcp.R900001-MCP200] [PMID: 19955089]

[74]   Ong SE, Blagoev B, Kratchmarova I, *et al.* Stable isotope labeling by amino acids in cell culture, SILAC, as a simple and accurate approach to expression proteomics. Mol Cell Proteomics 2002; 1(5): 376-86.
[http://dx.doi.org/10.1074/mcp.M200025-MCP200] [PMID: 12118079]

[75]   Phillips NJ, Steichen CT, Schilling B, *et al.* Proteomic analysis of *Neisseria gonorrhoeae* biofilms shows shift to anaerobic respiration and changes in nutrient transport and outermembrane proteins. PLoS One 2012; 7(6)e38303
[http://dx.doi.org/10.1371/journal.pone.0038303] [PMID: 22701624]

[76]   Guina T, Wu M, Miller SI, *et al.* Proteomic analysis of *Pseudomonas aeruginosa* grown under magnesium limitation. J Am Soc Mass Spectrom 2003; 14(7): 742-51.
[http://dx.doi.org/10.1016/S1044-0305(03)00133-8] [PMID: 12837596]

[77]   Booy AT, Haddow JD, Ohlund LB, Hardie DB, Olafson RW. Application of isotope coded affinity tag (ICAT) analysis for the identification of differentially expressed proteins following infection of atlantic salmon (Salmo salar) with infectious hematopoietic necrosis virus (IHNV) or Renibacterium salmoninarum (BKD). J Proteome Res 2005; 4(2): 325-34.
[http://dx.doi.org/10.1021/pr049840t] [PMID: 15822907]

[78]   Abdallah C, Dumas-Gaudot E, Renaut J, Sergeant K. Gel-based and gel-free quantitative proteomics approaches at a glance. International journal of plant genomics 2012; 2012
[http://dx.doi.org/10.1155/2012/494572]

[79]   Suriyanarayanan T, Qingsong L, Kwang LT, Mun LY, Truong T, Seneviratne CJ. Quantitative proteomics of strong and weak biofilm formers of *Enterococcus faecalis* reveals novel regulators of biofilm formation. Mol Cell Proteomics 2018; 17(4): 643-54.
[http://dx.doi.org/10.1074/mcp.RA117.000461] [PMID: 29358339]

[80]   Zhou X, Xing X, Hou J, Liu J. Quantitative proteomics analysis of proteins involved in alkane uptake comparing the profiling of *Pseudomonas aeruginosa* SJTD-1 in response to n-octadecane and n-hexadecane. PLoS One 2017; 12(6)e0179842
[http://dx.doi.org/10.1371/journal.pone.0179842] [PMID: 28662172]

[81]   Thompson A, Schäfer J, Kuhn K, *et al.* Tandem mass tags: a novel quantification strategy for comparative analysis of complex protein mixtures by MS/MS. Anal Chem 2003; 75(8): 1895-904.
[http://dx.doi.org/10.1021/ac0262560] [PMID: 12713048]

[82]   Parker J, Zhu N, Zhu M, Chen S. Profiling thiol redox proteome using isotope tagging mass spectrometry. JoVE (Journal of Visualized Experiments) 2012 Mar; 24(61): e3766.
[http://dx.doi.org/10.3791/3766]

[83]   Parker CE, Borchers CH. Mass spectrometry based biomarker discovery, verification, and validation--quality assurance and control of protein biomarker assays. Mol Oncol 2014; 8(4): 840-58.
[http://dx.doi.org/10.1016/j.molonc.2014.03.006] [PMID: 24713096]

[84]   Kellermann J. ICPL—isotope-coded protein label In 2D PAGE: Sample Preparation and Fractionation. Humana Press 2008; pp. 113-23.
[http://dx.doi.org/10.1007/978-1-60327-064-9_10]

[85]   Kellermann J, Lottspeich F. Isotope-coded protein label.Quantitative Methods in Proteomics. Totowa, NJ: Humana Press 2012; pp. 143-53.
[http://dx.doi.org/10.1007/978-1-61779-885-6_11]

[86]   Schmidt A, Kellermann J, Lottspeich F. A novel strategy for quantitative proteomics using isotope-coded protein labels. Proteomics 2005; 5(1): 4-15.
[http://dx.doi.org/10.1002/pmic.200400873] [PMID: 15602776]

[87]   Glenn WS, Stone SE, Ho SH, *et al.* Bioorthogonal Noncanonical Amino Acid Tagging (BONCAT) Enables Time-Resolved Analysis of Protein Synthesis in Native Plant Tissue. Plant Physiol 2017;

173(3): 1543-53.
[http://dx.doi.org/10.1104/pp.16.01762] [PMID: 28104718]

[88]　Landgraf P, Antileo ER, Schuman EM, Dieterich DC. BONCAT: metabolic labeling, click chemistry, and affinity purification of newly synthesized proteomes. Methods Mol Biol 2015; 1266: 199-215.
[http://dx.doi.org/10.1007/978-1-4939-2272-7_14] [PMID: 25560077]

CHAPTER 2

Past, Present, and Future of Gel-Based Microbial Proteomics

Munna Lal Yadav[1,*], Arvind K. Verma[1], Preeti Rawat[1], Abhishek Parashar[1], Divakar Sharma[2], Sudarshan Kumar[1] and Ashok K. Mohanty[1,*]

[1] *Animal Biotechnology Centre, National Dairy Research Institute, Karnal, India*

[2] *Kusuma School of Biological Sciences, Indian Institute of Technology, Delhi, India*

Abstract: Proteome deals with complete set of proteins, expressed by a genome of a cell present inside of an organism at a specific time period. It includes expression study, information of modifications in proteins, interactions with other protein biomolecules, *etc.* The goal of proteomics is to know new protein information, whether a protein is over expressed or under expressed in specific situation and overall gives information of its effect on an organism. Thousands of proteins can be examined at a time as compared to other methodologies. Many advanced technologies have been evolved to investigate proteome in depth and generate a huge amount of data. The most common high throughput techniques such as polyacrylamide & agarose gel electro-phoresis and their advanced versions in combination with mass spectrometry are being used in modern proteomics. The study of microbial proteome data helps us to know how bacteria get resistant to particular drug and which biomolecules are involved in that process. Database also gives the opportunity to develop better drugs that target new places on bacterial surface or new drugs on the same target. The advancement of proteomics technique and their applications in microbial research has granted a new hope to explore disease biomarkers and the development of diagnostic assays. Microbial protein's benefits are extensively used in the agricultural sector. Proteomics profiling has a key role in disease identification in humans and animals. Thanks to protein information as new targets are identified and more safer and effective drugs are produced.

Keywords: 2D PAGE, Biomarker discovery, Dairy processing, DeCyder, DIGE, Gel electrophoresis, IEF, Microbial proteomics, SDD-AGE, SDS-PAGE, Proteome, Tricine.

[*] **Corresponding author Munna Lal Yadav:** Animal Biotechnology Centre, National Dairy Research Institute, Karnal, India; Tel: 0184-2259542; E-mails: munna.biotek@gmail.com and ashokmohanty1@gmail.com

Divakar Sharma (Ed.)
All rights reserved-© 2020 Bentham Science Publishers

INTRODUCTION

Proteomics is the branch of life science that deals with the study of protein expression, protein modifications and interactions between proteins. It involves the identification of proteins with a possible role, change in protein expression, modification such as glycosylation, phosphorylation and interaction with other molecules inside the cell [1 - 3]. Proteomics has major implications in understanding the physiological process better. This allows the researcher to identify various proteins that are expressed inside a cell, which may not be the same as the transcriptome profile [2]. Generation and analysis of large proteomics data are expected to show how cells vigorously respond to changes in their surroundings, and disease conditions [2, 4].

Microbial proteome helps us to know how microbes get resistant to a particular drug, their biochemical pathway involves. This information is vital in selecting/creating novel antibiotics or drugs for their destruction [5, 6]. Further, it gives new hope to explore disease biomarkers and development of diagnostic assays. On the other hand, profiling of good bacteria proteins is important for the agriculture sector for increasing yield of crops and dairy processing [5, 7 - 8]. Many advanced technologies have been evolved to investigate proteome in-depth and generate a huge amount of data. The most commonly high throughput techniques used are gel based methods: Polyacrylamide & agarose Gel Electrophoresis and their advanced versions (SDS-PAGE, Native PAGE, SDD-AGE, Tricine SDS-PAGE, 2D-PAGE, DIGE and mass spectrometry (MS). 2D-PAGE or with LC-MS (Liquid Chromatography–Mass Spectrometry) are used in comparative proteomics study such as difference profiling of protein from normal condition to disease condition [2].

PROTEIN SAMPLE PREPARATION FROM MICROBIAL SOURCE

Microbial cell wall or cell membranes are different from other mammalian cells or tissues. Bacterial cell wall is relatively rigid and tough. It is important to have an idea about the localization of proteins. As some of the proteins are secreted in medium so their presence needs tobe checked in medium instead of cell lysate fraction [9]. The detergents are considered good to break cell wall proteins. Lysozyme is another component of bacterial lysis buffer which breaks peptidoglycan bonds. Some non-ionic detergent solutions are commercially available that mimic the role of lysozyme. The combination of DNase I and Lysozyme gives good result in purifying large molecular weight proteins [4, 10].

For a comprehensive analysis of all intracellular proteins, the cells should be efficiently disrupted. During protein extraction method, it is important to maintain temperature throughout cell lysis process and centrifugation. Proteases maybe liberated upon cell disruption. Cell lysis leads to the action of unregulated action of endogenous protease and phosphatase on the cellular protein [10, 11]. Over time the actual activity of the target protein in the study or whole cellular protein concentration is reduced.To minimize these effects and acquire the simplest possible protein yield after cell lysis, protease inhibitor cocktails are added to the lysis buffer [4, 12].

The protocols for proteome analysis depends upon various factors such as sample type, its physical and chemical properties, location of proteins inside the cell, availability of quantity and designing of an experiment [2, 5]. The methods are optimized at different steps for accurate quantification of proteins and comparison without loss of the actual sample. Sample must be prepared in such a way that there is minimum loss of proteins and various proteins are maintained in their true state of expression. The main challenges are the small sample size, difficulties in extraction, losses associated with protein precipitation and preservation of proteins without loss of low abundant proteins [2, 4 - 5, 13]. Further, for the solubilization of proteins, compatible buffers are required. In 2D gel electrophoresis urea is the choice of chemical for protein solubilization, however, in shotgun applications, most of the proteins are solubilized in ammonium bicarbonate containing buffers.

CLEAN-UP AND QUANTIFICATION OF PROTEINS

Protein sample clean up refers to the process of removing non-protein agents (detergents, salts and other molecules used in or generated during protein extraction or purification) that can hinder the results of downstream processes such as isoelectric focusing (IEF), 2D electrophoresis, adverse effects on protein function or stability, or interfere with downstream applications [4, 11]. Therefore, it may be necessary to remove or reduce these contaminants by different methods of protein precipitations (ammonium sulphate, acetone, trichloro acetic acid) or dialysis and desalting of the samples. Quantification of protein requires before proceeding to isolation, purification, and gel analysis [14 - 15]. Most protein estimation methods depend on the levels of tryptophan, tyrosine and other aromatic amino acids. There are many methods of protein estimation but Bradford assay and Bicinchoninic Acid (BCA) assay are most widely used.

GEL ELECTROPHORESIS OF PROTEINS

Gel Electrophoresis is a process where macromolecules like proteins and nucleic acids are moved on the gel in the presence of electric current. The rate of movement depends upon several factors such as a charge on the molecules, size, intensity of electric current, and temperature [1, 16 - 17]. The methods can be divided into two types: (1) Agarose gel,(2) Polyacrylamide gel (PAGE).

1. Agarose is a linear polysaccharide that comes from seaweeds. It is made of repeating units of agarobiose. The percentage of agarose in gel determines the pore size of gel. If the percentage is low then large size molecules can pass through the gel. However, at a higher percentage the trend is reversed. Larger biomolecules such as nucleic acids and some protein complexes are separated in agarose gel [16].
2. Polyacrylamide is made of polymerization of the acrylamide monomer which are cross linked by 'bis'- acrylamide. Ammonium persulfate (APS) and TEMED (-N, N, N, N Tetramethylethylene-1, 2-diamine) are required for free radicals generation, a key process in the polymerization of acrylamide monomer [16, 17].

TYPES OF PAGE

SDS-PAGE (Sodium Dodecyl Sulphate-polyacrylamide Gel Electrophoresis)

SDS-PAGE is basically used in the separation of proteins on the basis of their size and molecular mass. It is the choice of method for protein purification and determining the overall molecular mass of the proteins. SDS is an anionic detergent that has an overall negative charge, as most of the proteins have a positive and negative charge, it is better to give a protein overall charge. In this method, SDS binds to water repelling amino acids of proteins and gives an overall negative charge to the proteins. SDS is also responsible for denaturing the proteins. Therefore, proteins are separated in SDS-PAGE on the basis of size [17]. The method uses two type of gels, stacking gels and separating gels. The stacking gel has a large pore size with pH in the range of 7. Its main role is to concentrate the proteins under the influence of an electric current. Due to the concentration of the proteins, a sharp band is created that ultimately separates the proteins in separating gel. The separating gel is also known as resolving gel, it contains a higher percentage of acrylamide as compared to stacking gel and its pH is also basic in nature. Here, proteins are separated according to their size (Figs. **1** and **2**) [17 - 19].

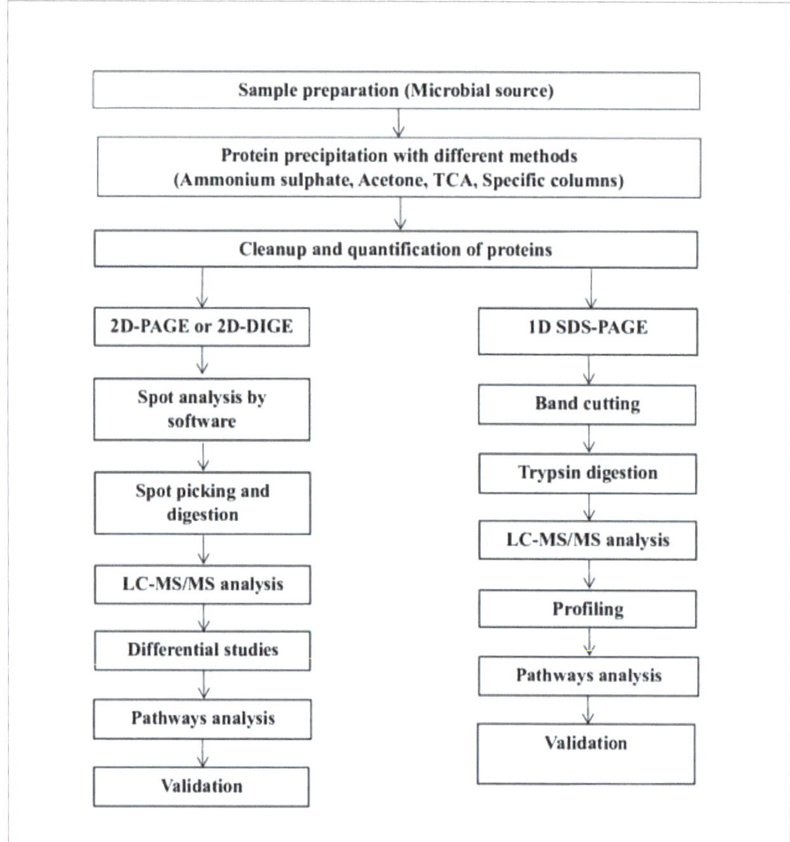

Fig. (1). An outline of gel based microbial proteomics and biomarker discovery.

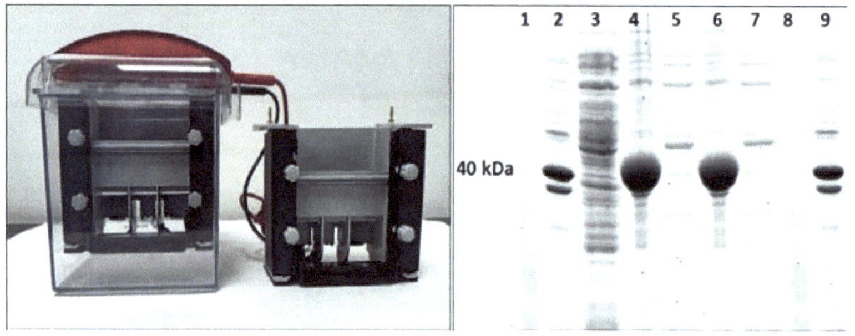

Fig. (2). (A) SDS-PAGE Unit (Courtesy from GE Healthcare) and **(B)** SDS-PAGE of proteins (MW of 40 kDa).

Tricine SDS-PAGE

Tricine SDS-PAGE is employed to separate proteins and peptides starting from 1 to 50 kDa molecular weights [20, 21]. The method uses low percentage of acrylamide as compared to other variant of PAGE. The conventional SDS-PAGE uses glycine for the separation of proteins, however, this method uses tricine. The low pK value of tricine facilitates the separation of low molecular weight proteins and peptides. The method is also fast compared to other PAGE methods and widely used for further analysis of proteins in LC-MS [21].

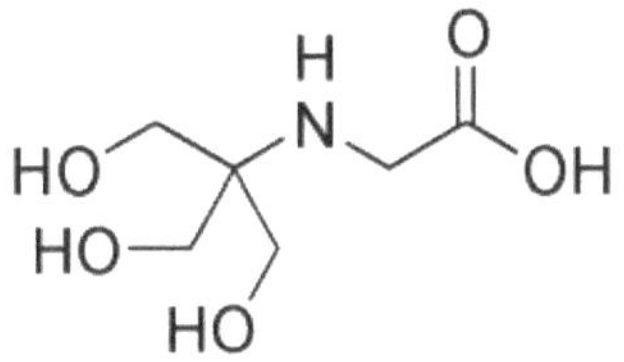

Tricine (N-(2-Hydroxy-1, 1-bis (hydroxymethyl) ethyl) glycine).

Native Page

Native PAGE is also a type of gel electrophoresis method for separating proteins in their native conditions. The proteins are not denatured here as done in SDS-PAGE methods. Proteins gets separated on the basis of their charge and size/confirmation. Compact structure proteins have higher mobility as compared to unfolded structure of proteins. The running condition of Native PAGE is always neutral in nature [22]. The main advantage using native PAGE is that it gives proper information whether the isolated protein is in its native form or degraded form. The method is very useful in knowing protein protein interactions and protein interaction with other molecules.

SDD-AGE (Semi Denaturing Detergent-agarose Gel Electrophoresis)

SDD-AGE (Semi Denaturating Detergent Agarose Gel Electrophoresis) is another type of protein separating method. The method does not use polyacrylamide, instead uses agarose with SDS. Because of large size of agarose it allows large protein complexes such as ameloid and prions to pass through separating method [11]. SDS-AGE does not completely denature protein, care should be taken that the proteolytic enzymes does not fully degrade desired protein.

2D-PAGE (Two-dimensional Gel Electrophoresis)

Two-Dimensional gel electrophoresis (2D-PAGE) exploits two independent character of proteins. Here, proteins are separated on the basis of isoelectric points (pIs) and their molecular weights (MW) [1, 16 - 17]. The procedure starts with IEF and ends with SDS-PAGE. Proteins separated in 2D gels are visualized by conventional staining method, including Coomassie staining, silver staining and fluorescent dye (Fig. **3**). Silver and fluorescent dyes are the most sensitive staining methods in comparison to Coomassie [2, 10].

The main application of 2D gel electrophoresis is identifying post and co-translational modifications in proteins. The method is very helpful in detecting marker proteins associated with antibiotic resistant, abnormalities such as cancer *etc* [1, 7 - 9, 10, 13, 15, 23 - 30].

Isoelectric Focusing (First Dimension)

In Isoelectric Focusing (IEF), proteins get separated on the basis of their isoelectric point. The IEF of a protein is pH value at which it has no net charge or in other words, protein contains an equal number of cations and anions. If the pH value of a protein is higher than its pI value, the protein is considered to be negatively charged and moves towards a positive electrode. However, if the pH value is lower than the pI value of protein, then protein behaves as a positively charged molecule. The method uses ampholytes, which contains pH gradient in the gel. As the proteins are added on the gel, they get separated on the basis of their pIs value. Phosphorylated proteins are one of the examples that are separated in IEF electrophoresis [14].

Fig. (3). Isoelectric Focusing (IEF) system (GE healthcare).

The immobilized pH gradient (IPG) is generated by covalently incorporating a gradient of immobilines (acid and basic buffering groups) at the time of polyacrylamide gel casting. The general structure of immobiline reagent is:

$$\textbf{CH2 CH–C–NH–R}$$

$$\textbf{||}$$

$$\textbf{O}$$

R = weakly acidic or basic buffering group

SDS-PAGE (Second Dimension)

The second dimension SDS-PAGE is performed on numerous vertical systems after completion of IEF. In SDS-PAGE, proteins get separated according to their molecular size and weights. It is an extensively used method for qualitative and quantitative analysis of protein mixture. The detailed role of SDS in protein separation has been mentioned in SDS-PAGE section.

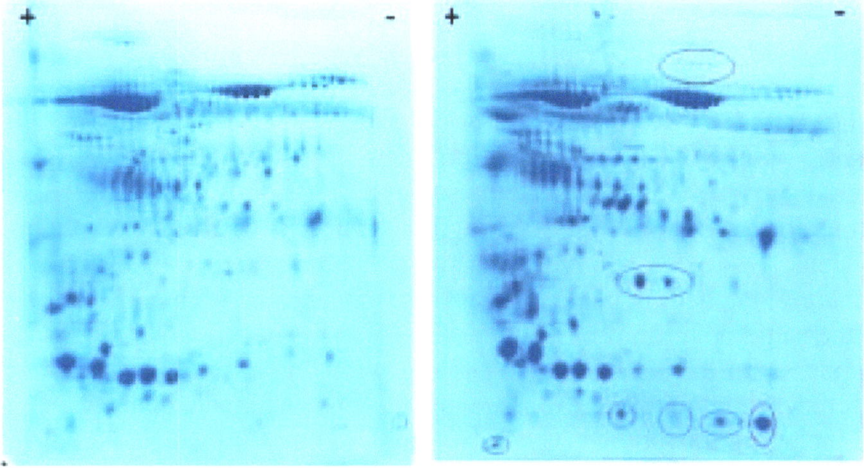

Fig. (4). Two dimensional gel electrophoresis (2D-PAGE) image.

SDS-PAGE Procedure Steps After Completion of IEF

1. Preparation of the system for protein separation in second- dimension electrophoresis.
2. Immobiline strips are equilibrated with equilibration buffer containing urea, SDS, Tris, glycerol and DTT.
3. Placing the equilibrated IPG strips on the separating gel of the SDS-PAGE.
4. Running the gel electrophoresis at low voltage for 2nd dimension separation.

2D-DIGE (TWO DIMENSIONAL DIFFERENCE IN-GEL ELECTROPHORESIS)

Difference in gel electrophoresis (DIGE) has become a powerful quantitative, sensitive and reliable technology for differential-display proteomics with the individual abundance changes for thousands of intact proteins that can be simultaneously monitored in replicate samples over multiple variables with statistical confidence [15, 31 - 32]. This technology is widely used enabling the detection of more refined changes in protein expression than classical 2D gel electrophoresis. The DIGE technique was first described at Jon Minden's laboratory [5], which has subsequently been refined and marketed by GE Healthcare Life Sciences (http://www.ettandige.com).The technology is based on the specific properties of spectrally resolvable fluorescent dyes (CyDye DIGE Fluor dyes). In this method, protein samples are labelled with cyanine fluorescent dyes (Cy2, Cy3, and Cy5) prior to 2-D electrophoresis [5, 31, 33]. Two sets of dyes are there, such as minimal dyes (Cy2, Cy3, and Cy5), and saturation dyes (Cy3 and Cy5). The minimal dyes are used where sufficient amounts of sample are available. Minimal labelling is the more established DIGE chemistry, where all three dyes can be used at a low stoichiometry of labelling to avoid protein precipitation owing to the prevalence of lysines in proteins and the hydrophobic nature of the dyes [33 - 34]. The saturation dyes are used in the cases where sample quantity is very less. The saturation labeling utilizes maleimide reagents for the stoichiometric labelling of cysteine sulfhydryls.The maliemide dyes can be used for stoichiometric labelling with minimal concerns of dye hydrophobicity because cysteine residues are much less prevalent in proteins. These dyes are mass and charge-matched, hence identical proteins labelled with each of the dye migrate to the same position on a 2-D gel [5, 32]. This ability allows separating more than one sample inindividual gel along with the internal reference. The internal reference (standard) is prepared by mixing an equal quantity of each sample and including this pooled (Cy2, Cy3 and Cy5) mixture on each gel [5, 15, 31, 34].

The DIGE System includes a defined method for 2-D that uses an internal standard to get the foremost correct results. It is a dominant tool for separating complex mixtures of proteins by charge and size and analyzing the second dimension SDS PAGE gel images for protein differences (Fig. **5**) [5, 31]. The multiplexing ability of the 2D DIGE methodology allows the combination of the same internal standard on each gel and thereby eliminates gel to gel variation. Ensuring that each and every protein spot has its internal standard is the only way to remove gel to gel variation, thereby significantly increasing accuracy and reproducibility [33]. To take advantage of this ability to multiplex, DeCyder 2D software and Image Master 2D Platinum softwareare used to precisely measure very small protein differences with high confidence (Fig. **6**) [32, 34 - 35]. This software contains proprietary algorithms that perform co-recognition of differently labelled samples within same gel. The short listed protein spots are picked from the gel for mass spectrometry [15, 31].

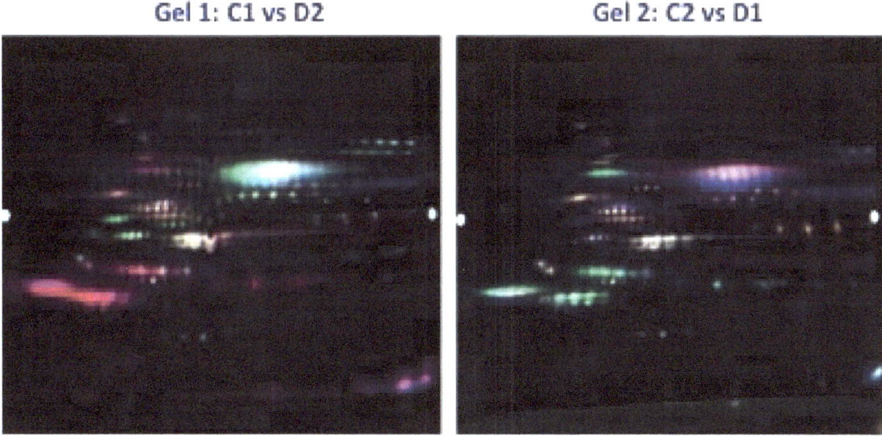

Fig. (5). Differential In Gel Electrophoresis (DIGE) image.

DIGE is a powerful quantitative technology within the repertoire of proteomics techniques for therapeutic target and biomarker discovery [15, 31]. Despite being one of the best protein profiling technologies and well-suited to quantitative measurements from the extended cohort, it has yet to fulfil its promise as a high throughput technology capable of processing the sizes of patient cohort necessary for the discovery of diagnostic and prognostic protein biomarkers. The use of the 2D DIGE/MS technology has the potential to make a significant impact within clinical diagnoses and treatment with the use of careful experimental design, sample simplification and robust statistical analyses [5, 15, 24, 32].

ANALYSIS OF 2D GEL AND DIGE GEL BY DIFFERENT SOFTWARE

2D-PAGEand DIGE gel analysis softwares are employed in gel electrophoresis to investigate bio-markers by quantifying individual proteins, to point out the separation between one or more protein spots on a scanned image of 2D gel or DIGE gel [15, 31]. The matching of spots between gels of similar samples gives an idea of protcomic differences between normal and advanced stages of unwellness. The 2D gel analysis usually addresses questions like protein level differences caused by a life-cycle stage, disease state and drug treatment *etc*. (Figs. **4** and **5**). There are many softwares (ImageMaster Platinum and DeCyder) that are used to accurately measure the protein differences with high confidence (Fig. **6**) [15, 31, 34 - 35].

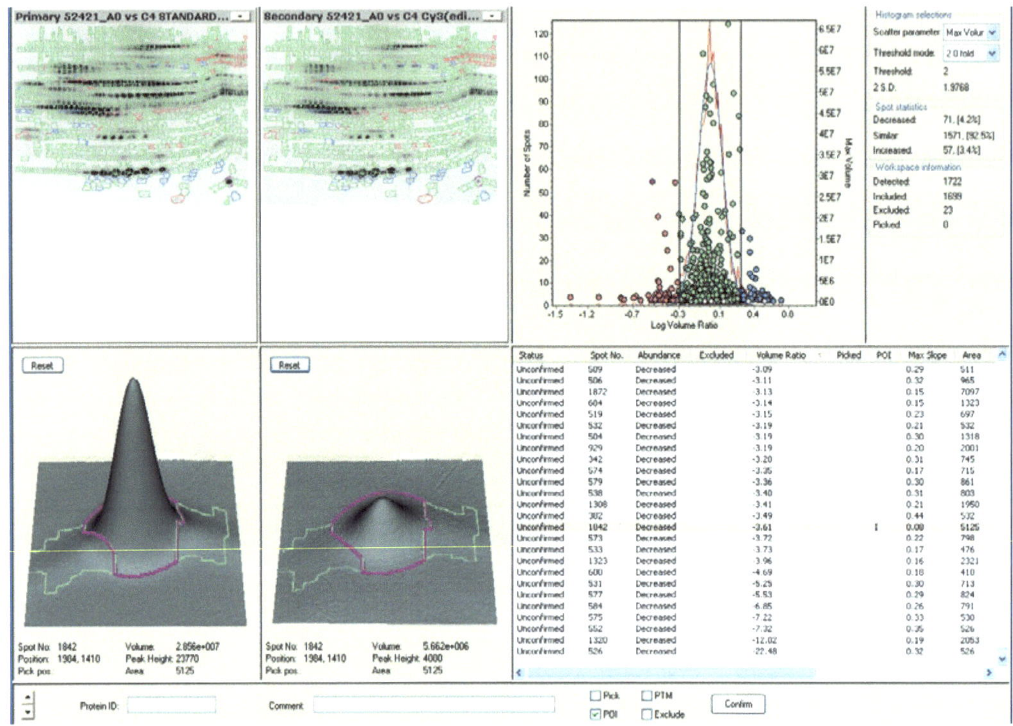

Fig. (6). Differential In-Gel Analysis (DIA) by DeCyder.

2D Gel Analysis using Image Master 2D Platinum

Image Master 2D Platinum is a flexible software for the inclusive visualization, examination, and analysis of 2-D gel data from both 2-D DIGE (Fluorescence Difference Gel Electrophoresis) and non-DIGE methodologies [10, 36]. ImageMaster 2D Platinum, powered by the innovative and widely used Melanie gel analysis software developed by the Swiss Institute of Bioinformatics in association with GeneBio and GE Healthcare [15, 34]. It is simplified and enrich the analyses of 2-D gels and helps in the identification of protein biomarkers. Several the functions found in previous versions have been improved and simplified, and significant new features have been added for more suitable and easier working. The user interface has been upgraded and new interaction modes simplify even the most complex analysis [10, 15, 31, 36].

Evaluation of DIGE Gels using DeCyder

Difference gel electrophoresis (DIGE) gels can be assessed with several distinct algorithms. However, in order to get the full use of all features of the DIGE concept, it is highly recommended to apply the DeCyder 2-D software package, which is highly suited to this type of application, enabling it to detect and quantitate complex changes in protein expression. Scanning of 2D gel is performed with Typhoon Variable Mode Imager, which is a more sensitive imager adapted to meet the specific requirements of 2-D DIGE. The system provides different imaging modes such as fluorescence, filmless autoradiography and chemiluminescence detection. The system is optimized for differential protein studies and quantitative protein detection. Imager system is equipped with green 532 nm (Cy3), blue 473 nm (Cy2) and red 650 nm (Cy5) lasers, which optimally detects Cy3, Cy2, and Cy5 signals. It exhibits outstanding linearity and quantitative accuracy. By generating overlaid, multichannel images for each gel with minimal crosstalk, typhoon exploits the multiplexing potential of CyDye DIGE fluorescents to remove experimental variation between gels. For the detection of radioactivity and fluorescence, light emitted from a fluorescently labeled sample is collected and transformed into an electrical signal by a photomultiplier tube. The electrical signal is then converted into digital information for image analysis. The images are analyzed using Decyder 2D software to accurately measure small differences in protein. Typhoon variable mode imager, permits creation of quantitative and reproducible data, and offers a sensitivity lower to 125 pg proteins compared to 5 ng for silver staining [15, 31, 34 - 35].

Image analysis is done with DeCyder 2-D Differential Analysis Software. It is an automated image analysis software, which is used for the analysis of differentially expressed proteins of 2D Fluorescence DIGE experiments [34]. For analysis by Decyder, Differential In-gel Analysis (DIA) and the Biological Variation Analysis (BVA) modules: The DIA module generates the information of all the protein spots present on each gel. This information is then used for inter gel matching. DIA module process a pair of images from a single gel. BVA performs quantitation of protein expression across multiple gels. Spot detection is based on the initial detection of protein spots in the internal standard (Cy2) image of pooled samples, which is followed by the application of similar boundaries to the remaining images (Cy3 and Cy5) within each gel. Since the internal standard (Cy2) was identical on all gels, the software performed the matching only on the internal standard images labeled with Cy2. The gel with the maximum number of protein spot was selected as a master gel. The existence of the reference sample on each gel allows accurate normalization of every samples [15]. This permits direct spot volume ratio measurements and produces a precise comparison of every protein with its representative internal standard [32, 35]. The software spontaneously analyze detection, quantitation, normalization and background subtraction, which takes into account any differences in the dyes, *i.e.* molar extinction coefficient, quantum yields, *etc.* During analysis, differences in expression of less than 10%, with over 95% confidence, can be attained within minutes. In combination with CyDye DIGE Fluor dyes, DeCyder software allows gel analysis using different experimental designs with several degrees of complexity [5, 31]. A simple control, treated experiment through to a multi-factorial experiment addressing factors such as dose and time, can be accomplished in a single analysis [33, 35].

COMPARISON BETWEEN IEF, SDS-PAGE, 2D-PAGE, AND DIGE

Technique	IEF	SDS-PAGE	2D-PAGE	DIGE
Principles	Isoelectric focusing (IEF) is an electrophoretic technique for the first dimension separation of proteins based on their isoelectric point (pI).	SDS-PAGE is an electrophoretic method in which proteins get separated according to their molecular weights.	2D PAGE is a combined method including IEF and SDS-PAGE.	DIGE is a combined method of IEF and SDS-PAGE using cyanine fluorescent dyes (Cy2, Cy3, and Cy5) labelled samples.
pH gradient	The IEF involves placing the sample in a gel with a different pH gradient.	pH gradient not required	The first dimension needs pH gradient and proteins get separated in the second-dimension on the basis of their molecular mass.	First dimension needs a pH gradient and proteins get separated in the second-dimension on the basis of their molecular mass.
Ampholites	Ampholites of different pH range required in IEF.	No ampholyte requirement	Required in first dimension protein separation	Required in the first dimension
Migration of Proteins	Horizontal	Vertical	Both dimension	Both dimension
Separation of proteins	Low proficiency of separation.	Slightly high capability of separation.	High efficiency for separation of hundreds of Proteins.	Very high capability for separation of thousands of Proteins in a single gel run.
Capability of detection	Individually low	Individually low	High Capability to Detect post translational modifications also	Very high Capability to detect thousands of proteins up to three samples run in a single gel.

Technique	IEF	SDS-PAGE	2D-PAGE	DIGE
Handling and running cost	Ease to perform and relatively low in cost	Ease to perform and relatively low in cost	Ease to perform and relatively low in cost	Expertise needed to perform and analysis of results, relatively high in cost

MICROBIAL PROTEOMICS IN DAIRY PROCESSING

Many bacterial species are widely used in the dairy industry for the production of quality dairy products. Milk processing procedures are extremely dependent on the total range of milk proteins. The bacterial presence in milk may change the quality of milk that might significantly affect the production processes. Bacterial behavior itself changes drastically when it is moved from the research laboratory to the manufacturing domain and challenges the production strategies [2, 6]. Microbial proteomics can give important information regarding changes in protein expression of a bacteria. These changes are key indicators of the adaptation of microorganisms in different conditions [37]. The proteomes of many lactic acid bacteria have been studied, such as *Lactobacillus bulgaricus*, *Lactococcus lactis*, and *Streptococcus thermophiles*. Proteome data of these bacteria gives valuable information in modifying or selecting specific bacteria for good quality dairy food making.

Probiotics are right now one of the potential research areas of food biotechnology, where gut friendly bacteria are used to keep the human gut healthy. However, observing the number of their applications in the dairy industry and human health, it is felt that screening and identification of probiotics are required. The recent developments and innovations in proteomics will allow rapid characterization of a huge number of potential bacterial species as probiotics. Proteome data for probiotics and pathogenic bacteria can be generated to improve quality milk products and its processing strategies in the near future [2, 19].

FUTURE PERSPECTIVES OF GEL BASED MICROBIAL PROTEOMICS

Proteomics has major implications in the understanding of the physiological process. This allows the researcher to identify various proteins that are expressed inside a cell, which may not be the same as the transcriptomic profile [7, 15, 23, 31]. The major challenge in the field of proteomics includes an incomplete understanding of the protein function of a given organism. Most of the methods for protein analysis are cumbersome to perform. Easy and quick methods are needed. Further, it is difficult to predict protein function, interaction with other bimolecular in real time, as most of the protein studied get denatured during isolation, purification *etc* [17]. Despite shortcomings, new methods help in unleashing new information on protein functions, their behavior and their role in diseases [2, 4, 7 - 9, 13, 15, 19, 23 - 30, 33].

Like any other biological system, proteomics can be applied to microbial science, to complement genomics. Some of the aspects which can be thought of in microbial proteomics are:

1. Building proteome map of different pathogenic and non-pathogenic strains of microbes. Microbial proteome project can be initiated for cataloguing various proteins in bacteria, fungi and virus in line with human proteome project.
2. Differential expression proteomics will help to quantitate the expression of various proteins in different species/strains for identification of protein biomarkers, and their pathogenic and beneficial effects can be assessed by functional genomics studies.
3. Pathogenesis promoting attributes of various microbial proteins can be assessed by proteomics studies.
4. Climate change effects can be studied in terms of the expression of bad and good proteins and their future application as biomarkers of stress.
5. Proteome data for probiotics and lactic acid bacteria can be generated to produce quality milk products, and its processing in the dairy industry.

CONSENT FOR PUBLICATION

Not Applicable.

CONFLICT OF INTEREST

The authors declare no conflict of interest, financial or otherwise.

ACKNOWLEDGEMENTS

The authors sincerely thank NDRI Karnal for providing laboratory and other facilities.

REFERENCES

[1]   Wilson K, Walker J, Eds. Principles and techniques of biochemistry and molecular biology. Cambridge university press 2010.
[http://dx.doi.org/10.1017/CBO9780511841477]

[2]   Mohanty AK. Introduction to proteomics and mass spectrometry for application in dairy processing and food safety. compendium: Advance Techniques and Novel Approaches for Quality and Safety Evaluation of Dairy Foods, Compendium. 2015.

[3]   Singh B, Mal G, Gautam SK, Mukesh M. Proteomics: Applications in Livestock. Advances in Animal Biotechnology. Cham: Springer 2019.

[http://dx.doi.org/10.1007/978-3-030-21309-1_34]

[4]   Cull M, McHenry CS. Preparation of extracts from prokaryotes. Methods in enzymology Academic Press. 1990; 182: pp. 147-53.

[5]   Lilley KS, Friedman DB. Difference gel electrophoresis DIGE. Drug Discov Today Technol 2006; 3(3): 347-53.
[http://dx.doi.org/10.1016/j.ddtec.2006.09.013] [PMID: 24980539]

[6]   Toscano M, de Grandi R, Drago L. Proteomics: the new era of microbiology. MicrobiologiaMedica 2017; 32: 4.

[7]   Lata M, Sharma D, Deo N, Tiwari PK, Bisht D, Venkatesan K. Proteomic analysis of ofloxacin-mono resistant *Mycobacterium tuberculosis* isolates. J Proteomics 2015; 127(Pt A): 114-21.
[http://dx.doi.org/10.1016/j.jprot.2015.07.031] [PMID: 26238929]

[8]   Khan A, Sharma D, Faheem M, Bisht D, Khan AU. Proteomic analysis of a carbapenem-resistant Klebsiella pneumoniae strain in response to meropenem stress. J Glob Antimicrob Resist 2017; 8: 172-8.
[http://dx.doi.org/10.1016/j.jgar.2016.12.010] [PMID: 28219823]

[9]   Sharma D, Bisht D. Secretory proteome analysis of streptomycin-resistant *Mycobacterium tuberculosis* clinical isolates. SLAS DISCOVERY: Advancing Life Sciences R & D 2017; 22: 1229-38.
[http://dx.doi.org/10.1177/2472555217698428]

[10]   Gorg A. Hand Book on 2-D Electrophoresis Principles and Methods, GE Health Care . 2004; pp. 80-6429-60.

[11]   Halfmann R, Lindquist S. Screening for amyloid aggregation by semi-denaturing detergent-agarose gel electrophoresis. J Vis Exp 2008; 16(17)e838
[http://dx.doi.org/10.3791/838] [PMID: 19066511]

[12]   Pergande MR, Cologna SM. Isoelectric point separations of peptides and proteins. Proteomes 2017; 5(1): 4.
[http://dx.doi.org/10.3390/proteomes5010004] [PMID: 28248255]

[13]   Sharma D, Bisht D. An efficient and rapid method for enrichment of lipophilic proteins from *Mycobacterium tuberculosis* H37Rv for two-dimensional gel electrophoresis. Electrophoresis 2016; 37(9): 1187-90.
[http://dx.doi.org/10.1002/elps.201600025] [PMID: 26935602]

[14]   Anderson JC, Peck SC. Detection of protein phosphorylation and charge isoforms using vertical one-dimensional isoelectric focusing gels InPlant MAP Kinases. New York, NY: Humana Press 2014; pp. 39-46.

[15]   Rawat P, Bathla S, Baithalu R, *et al.* Identification of potential protein biomarkers for early detection of pregnancy in cow urine using 2D DIGE and label free quantitation. Clin Proteomics 2016; 13: 15.
[http://dx.doi.org/10.1186/s12014-016-9116-y] [PMID: 27429603]

[16]   Mohanty AK, Yadav ML, Choudhary S. Gel electrophoresis of proteins and nucleic acids. Protocols in Semen Biology. Singapore: Springer 2017; pp. 233-46.
[http://dx.doi.org/10.1007/978-981-10-5200-2_18]

[17]   Sambrook J, Russell DW. Molecular cloning: a laboratory manual. 3rd edn.,

[18]   Laemmli UK. Cleavage of structural proteins during the assembly of the head of bacteriophage T4. Nature 1970; 227(5259): 680-5.
[http://dx.doi.org/10.1038/227680a0] [PMID: 5432063]

[19]   Walker JM, Ed. The proteomics protocols handbook. Humana Press 2005.
[http://dx.doi.org/10.1385/1592598900]

[20]   Schägger H. Tricine-SDS-PAGE. Nat Protoc 2006; 1(1): 16-22.

[http://dx.doi.org/10.1038/nprot.2006.4] [PMID: 17406207]

[21]   Jiang S, Liu S, Zhao C, Wu C. Developing protocols of tricine-SDS-PAGE for separation of polypeptides in the mass range 1-30 kDa with minigel electrophoresis system. Int J Electrochem Sci 2016; 11: 640-9.

[22]   Arndt C, Koristka S, Bartsch H, Bachmann M. Native polyacrylamide gels. Protein electrophoresis. Humana Press 2012; pp. 49-53.

[23]   Bathla S, Rawat P, Baithalu R, *et al.* Profiling of urinary proteins in Karan Fries cows reveals more than 1550 proteins. J Proteomics 2015; 127(Pt A): 193-201.
[http://dx.doi.org/10.1016/j.jprot.2015.05.026] [PMID: 26021477]

[24]   Monteoliva L, Albar JP. Differential proteomics: an overview of gel and non-gel based approaches. Brief Funct Genomics Proteomics 2004; 3(3): 220-39.
[http://dx.doi.org/10.1093/bfgp/3.3.220] [PMID: 15642186]

[25]   Sharma D, Kumar B, Lata M, *et al.* Comparative proteomic analysis of aminoglycosides resistant and susceptible *Mycobacterium tuberculosis* clinical isolates for exploring potential drug targets. PLoS One 2015; 10(10)e0139414
[http://dx.doi.org/10.1371/journal.pone.0139414] [PMID: 26436944]

[26]   Sharma D, Lata M, Singh R, Deo N, Venkatesan K, Bisht D. Cytosolic proteome profiling of aminoglycosides resistant *Mycobacterium tuberculosis* clinical isolates using MALDI-TOF/MS. Front Microbiol 2016; 7: 1816.
[http://dx.doi.org/10.3389/fmicb.2016.01816] [PMID: 27895634]

[27]   Lata M, Sharma D, Kumar B, *et al.* Proteome analysis of ofloxacin and moxifloxacin induced *Mycobacterium tuberculosis* isolates by proteomic approach. Protein Pept Lett 2015; 22(4): 362-71.
[http://dx.doi.org/10.2174/0929866522666150209113708] [PMID: 25666036]

[28]   Kumar B, Sharma D, Sharma P, Katoch VM, Venkatesan K, Bisht D. Proteomic analysis of *Mycobacterium tuberculosis* isolates resistant to kanamycin and amikacin. J Proteomics 2013; 94: 68-77.
[http://dx.doi.org/10.1016/j.jprot.2013.08.025] [PMID: 24036035]

[29]   Qayyum S, Sharma D, Bisht D, Khan AU. Identification of factors involved in *Enterococcus faecalis* biofilm under quercetin stress. Microb Pathog 2019; 126: 205-11.
[http://dx.doi.org/10.1016/j.micpath.2018.11.013] [PMID: 30423345]

[30]   Kumar G, Shankar H, Sharma D, *et al.* Proteomics of culture filtrate of prevalent *Mycobacterium tuberculosis* strains: 2D-PAGE map and MALDI-TOF/MS analysis. SLAS DISCOVERY: Advancing Life Sciences R&D 2017; 9: 1142-9.

[31]   Jena MK, Janjanam J, Naru J, *et al.* DIGE based proteome analysis of mammary gland tissue in water buffalo (Bubalus bubalis): lactating vis-a-vis heifer. J Proteomics 2015; 119: 100-11.
[http://dx.doi.org/10.1016/j.jprot.2015.01.018] [PMID: 25661041]

[32]   Unlü M, Morgan ME, Minden JS. Difference gel electrophoresis: a single gel method for detecting changes in protein extracts. Electrophoresis 1997; 18(11): 2071-7.
[http://dx.doi.org/10.1002/elps.1150181133] [PMID: 9420172]

[33]   Lee JE, Lee JY, Kim HR, Shin HY, Lin T, Jin DI. Proteomic Analysis of Bovine Pregnancy-specific Serum Proteins by 2D Fluorescence Difference Gel Electrophoresis. Asian-Australas J Anim Sci 2015; 28(6): 788-95.
[http://dx.doi.org/10.5713/ajas.14.0790] [PMID: 25925056]

[34]   Amersham Biosciences Ettan DIGE User manual. http:// www.amershambiosciences.com

[35]   Karp NA, Kreil DP, Lilley KS. Determining a significant change in protein expression with DeCyder during a pair-wise comparison using two-dimensional difference gel electrophoresis. Proteomics 2004; 4(5): 1421-32.
[http://dx.doi.org/10.1002/pmic.200300681] [PMID: 15188411]

[36]   https://pdf.medicalexpo.com/pdf/ge-healthcare-life-sciences/imagemaster-2d-platinum/-0554-165751.html

[37]   Chen B, Zhang D, Wang X, *et al.* Proteomics progresses in microbial physiology and clinical antimicrobial therapy. Eur J Clin Microbiol Infect Dis 2017; 36(3): 403-13.
[http://dx.doi.org/10.1007/s10096-016-2816-4] [PMID: 27812806]

CHAPTER 3

Liquid Chromatography-Mass Spectrometry (LCMS): An Advanced Tool for the Microbial Proteomics Analysis

Pranav Kumar Prabhakar[*]

Department of Medical Laboratory Sciences, Lovely Professional University, Punjab-144411, India

Abstract: Almost 20 years after the publication of the first microbial sequences, today's world is growing in the direction of the post-genomic era with the understanding of transcriptomics and proteomics, which offer insight into cellular physiology. As the function of protein is associated with the phenotypic characters of an organism, the power to exceed the entire protein network of a microbial world has a huge impact on microbiology. Nowadays, mass spectroscopy (MS) has been extensively used in microbiology for the identification, characterization, and serotyping. In recent times, the large genome size, its complexity, and their study is a big task for the microbiologist. It is interesting to see how scientists examine microbial proteomics and analyze them. Proteomics has a key task to carry out in endeavors to develop far-reaching cell maps of biochemical procedures happening inside explicit microorganisms at given spatial and worldly focuses. Here we are going to discuss the methodologies for the identification of bacteria up to species level with great accuracy with the use of proteomes of bacterial pure culture. This chapter will also discuss the sample preparation and identification of a specific strain of bacteria for the liquid chromatography-tandem mass spectrometry (LC-MS/MS) and the recent application of it.

Keywords: Bacteria, Chromatography, Electrospray, Isotope, Mass spectroscopy, Metabolic, Proteomics, Protein, Resistance, Transcriptomics.

INTRODUCTION

Several microbes grow in our surroundings and most of them are not visible with naked eyes. These microorganisms are extensively used for studies and research due to their simple cellular structure and less growth requirement. The genome of microbes is very small compared to most of the eukaryotes and hence easy to handle during the study, analyze, and sequencing. Proteomics has been efficiently

[*] **Corresponding author Pranav Kumar Prabhakar:** Department of Medical Laboratory Sciences, Lovely Professional University, Punjab-144411, India; Tel: 07696527883; E-mail: prabhakar.iitm@gmail.com

Divakar Sharma (Ed.)
All rights reserved-© 2020 Bentham Science Publishers

used for both fundamental and applied research to analyze and study cellular metabolic processes. There are several tools available in proteomics to probe the gene expression in the case of bacteria in given specific environmental conditions. The microbial proteomics has been used for the study of hotspot areas of choice such as stress responses, an adaptation of microbes in extreme environmental conditions, the pathophysiology of microbes, metabolic engineering. The ability of proteomics to address essential issues in the microbial field is largely subject to the supported advancement of various proteomic advances, which separately show their ability in proteomic exploration either subjectively or/and quantitatively [1].

The active role of biomolecules in the maintenance and helping in better life has been studied since the initial days of biochemical and biomedical researches. To explain the significance of these biomolecules in our life, in 1938, Berzelius used the term protein for them, which comes from the Greek and means the first [2]. All the protein content of a cell specified by its destination, protein-protein interaction, protein-DNA (Deoxyribose nucleic acid) interaction, posttranslational modifications, and its turnout rate is called the proteome. In 1996, Wilkins and Williams used the term proteomics first time for the total protein content of a cell, tissue, or organism and it denotes, "PROTein complement of a genOME" [3].

Most of the physiological information of DNA is expressed in terms of protein and is characterized by proteome. The proteome of prokaryotes is simple and easy to study compared to eukaryotes. The proteins expressed by prokaryotic bacterial cells are responsible for its pathogenicity and the study of such proteins isvery challenging and difficult to study due to huge differences in their physical and chemical properties such as amount produced, size and nature of protein, solubility, hydrophobicity, and lipophilicity, regulations [4].

Proteomics is a very effective tool for the early diagnosis of a disease, medical prognosis, and also to monitor the development and growth of disease as well as in the drug development process. All the proteome of a cell or tissue at any given time and in any given condition is characterized in terms of the proteomics, which includes its expression, molecular structure, shape and sizes, physiology and function, interactions with protein, or DNA, modification and post-transcriptional modifications [5]. The proteome of a cell, tissue, or organism varies from time to time and tissue to tissue or also in case of any kind of stimulations [6]. The variation in the gene expression can be studied and analyzed by the transcriptome and proteome to differentiate between the two physiological states [6].

Proteomics is one of the most significant techniques to encompass the gene function, though compared to genomics, it is more complex [7]. The microarray chip has been designed and developed for the large scale analysis of complete transcriptome [8]. Proteins of the cells are the functional component and their expression level is not only dependent on its mRNA (messenger ribose nucleic acid) level but also the translational regulatory system of the host cell and hence the proteomics is treated as one of the important data sets for analyzing the physiological status of a biological system [9]. There are several methodologies available for the proteome study. These techniques are classified into different subcategories such as conventional methods such as Chromatography, ELISA (Enzyme Linked Immuno Sorbent Assay), western blotting, advanced technologies (microarray-based techniques, gel-based assays, mass spectroscopy, sequencing), quantitative techniques, high throughput techniques, *etc.* The genome-based microarray is common, and the hybridization pattern of protein is completely different from the nucleic acid. The investigation of a huge number of proteins at a single time is very difficult as we require antibodies against each of the analyzed proteins and also the antibody binding pattern and process need to be optimized for each protein. And hence, protein microarrays are not a good choice for the complete proteome analysis process. Similar to the microarray, gel electrophoresis is also not a good choice to study gene expression as there is a limitation in the spotting, identification, and estimation and also not very effective for the mixture of proteins [10, 11]. Mass spectrometry (MS) has come with an efficient tool for the characterization of proteins from a complex mixture of protein [12]. There are several MS methods available for the investigation of the complete protein of cells/tissues. And they are "Surface-Enhanced Laser Desorption Ionization (SELDI)" [13], "Matrix-Assisted Laser Desorption Ionization (MALDI)" [14] "coupled with time-of-flight (TOF)" or other instruments, and "as chromatography MS (GC-MS)" or "liquid chromatography MS (LC-MS)." One of the limitations of SELDI and MALDI is that they do not induce on-line fractionation during MS analysis, and hence the protein complex mixture needs to be fractioned beforehand. MALDI is used in most of the cases for tissue imaging [15 - 17]. GS-MS or LC-MS has advantages over other MS tools that induce an online mode of separation for the protein test sample and hence commonly used for high-throughput proteomics.

None of these techniques are perfect as they are associated with some more drawbacks, although they also have advantages (Fig. **1**; Table **1**).

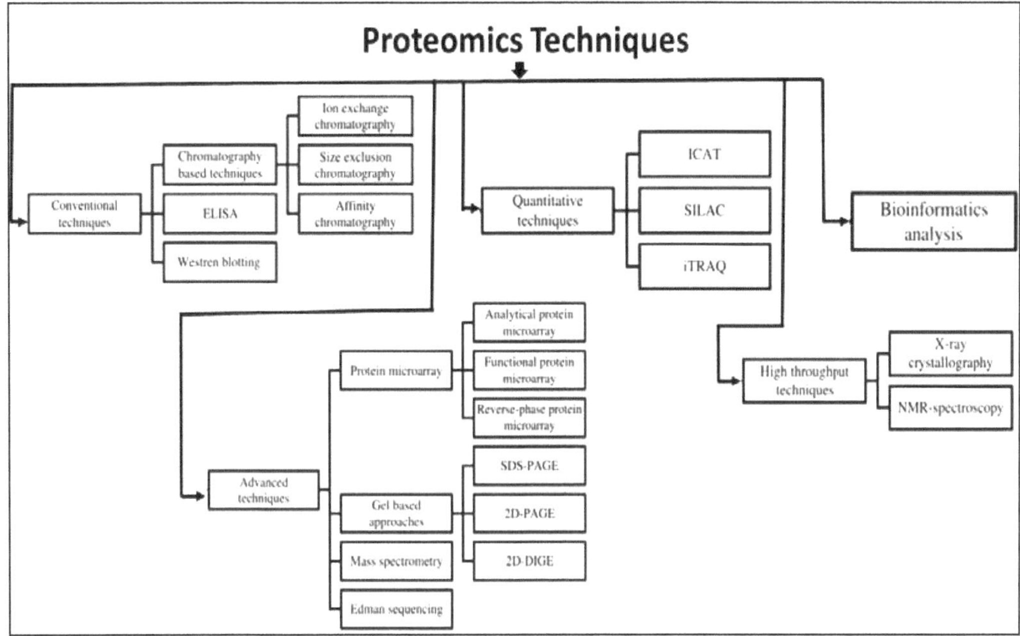

Fig. (1). Complete List of proteomics techniques.

Table 1. Advantages and disadvantages of proteomics analysis techniques.

Technology	Application	Strengths	Limitations
Ion Exchange Chromatography	Protein Purification	Low cost	Not good for proteome
Size-exclusion Chromatography	Protein Purification	Does not depend on temperature, pH, Solvent, ionic strength	Less selectivity and limited sample volume
Affinity Chromatography	Protein Purification	High level of purity and the reproducibility	ligand cost, low productivity, nonspecific adsorption, and degradation of solid support
"Isotope coded affinity tag (ICAT)"	Labeling of protein with isotopes	Very sensitive techniques, process is reproducible, even low expressed protein can be detected	Proteins and peptide lack cysteine cannot be detected; also not applicable for acidic proteins
"Stable isotope Labeling with an amino acid in cell culture (SILAC)"	Labeling of protein in cell culture and can quantify the expression level	High level of labeling and direct quantitation	Not applicable for tissue sample

(Table 1) cont.....

Technology	Application	Strengths	Limitations
"Isobaric tag for relative and absolute quantitation (iTRAQ)"	Isobaric labeling of peptides	High-throughput quantification and several samples can be quantified at one time	Sample complexity increases, Needs peptide fraction before mass spectroscopy
Protein array	Detect protein, biomarkers, enzymes on chip	More number of the sample can be used, very sensitive and require a very small amount of sample	Not very good for expression studies access a very large number of affinity reagents
Mass spectrometry	One of the primary tools for protein detection and identification	More number of the sample can be used, very sensitive, used for both qualitative and quantitative assay, also gives information about the post-translational modification	"No individual method to identify all proteins. Not sensitive enough to identify minor or weak spots. MALDI and ESI do not favor the identification of hydrophobic peptides and basic peptides"

MASS SPECTROSCOPY (MS)

In the last century, a study performed by J. J. Thomson and his fellow F. W. Aston discovered Mass Spectroscopy (MS) [18]. There are several tools available for molecular weight determination, but the MS has an advantage over the other available tools due to its high accuracy (~0.01–0.001%) and better sensitivity (detect as less as 10^{-9}-10^{-18} mol of the test sample) for the evaluation of the molecular weight of any biochemical compounds [19]. The instrument used is the mass spectrometer, which produces ion from the sample used and then isolates them based on mass by charge ration (m/z) and quantifies the relative occurrence of each ion to get a mass spectrogram [20]. The basic components of the mass spectrometer are an ion source that generates ion and ionizes proteins, a mass analyzer that separates peptide based on mass to charge ration and intensities, and a detector that detects these differentiated mass to charge ration for ionized peptide (Fig. **2**).

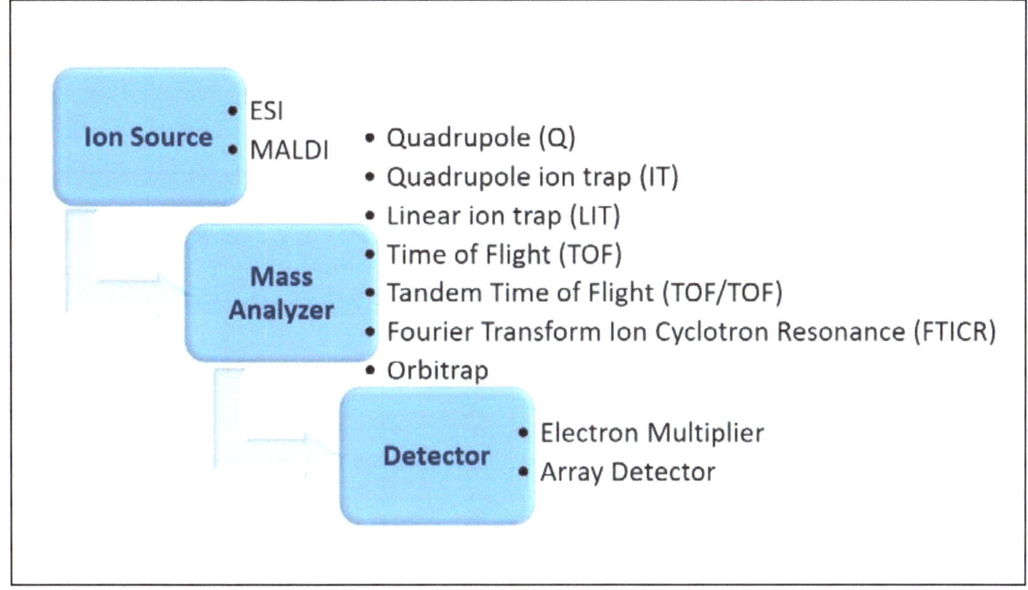

Fig. (2). Three basic components of mass spectrometry along with their examples.

Till now the mass spectroscopy was used for the estimation of molecular weight of relatively volatile compounds only, but after the soft ionization technique development by Fenn and his team (the 1980s), this technique is used for the very large, polar, and even thermolabile biomolecules like proteins and peptides, which was previously not possible [19 - 21]. Soft ionization is the ability in which the volatile thermolabile compounds are going to be ionized without splitting. Some of the examples of such thermolabile compounds are proteins [19]. The two important soft ionization technologies used for the characterization and quantification of protein are "electro spray ionization mass spectrometry (ESI-MS)" and "matrix-assisted laser desorption ionization-time of flight mass spectrometry (MALDI-TOF-MS)". Due to the significance of these techniques in biological and pharmaceutical science, Chemistry Nobel Prize (2002) was shared by John Fenn and Koichi Tanaka for their extraordinary work in the field of ESI and MALDI, respectively. MS is utilized to gauge the mass to charge proportion (m/z), in this way, supportive to decide the sub-atomic load of proteins. The general procedure contains three stages. The atoms must be changed to gas-stage particles in the initial step, which represents a test for biomolecules in a fluid or strong stage. The subsequent advance includes the detachment of particles based on m/z esteems within sight of electric or attractive fields in a compartment known as a mass analyzer. At long last, the isolated particles and the measure of every species with specific m/z esteem are estimated. Regularly utilized ionization technique involves matrix assisted laser desorption ionization (MALDI), surface improved laser desorption/Ionization (SELDI), and electro spray ionization (ESI) [22]. In clinical research centers, bacterial recognizable proof relies upon ordinary methods. In any case, distinguishing proof of moderate developing, critical, and anaerobic microscopic organisms through regular methods is costly, complex, and tedious. Biswas and Rolain [23] utilized the MALDI-TOF for early pathogenic bacteria distinguishing proof, which is helpful for early sickness control. MS has likewise turned into a noteworthy instrument in the detection of infection at the atomic level, and different infections and viral proteins, including unblemished infections, freak viral strains, capsid protein, post-translational changes were recognized [24]."

Shotgun MS-Based Proteomics

Due to the advancement made toward best in class cutting edge sequencing procedures applied to create completely clarified genomes, the quantity of completely sequenced microbial genomes has expanded drastically [25]. This has empowered best in class MS-based proteomics innovations to quickly progress to consider microbial models and their networks in a hearty and orderly manner. In this procedure, proteins are isolated by their sub-atomic weight and isoelectric point moving and collecting as protein spots on the gel network. Even though this methodology can resolve products of thousands of proteins, it carries a few inconveniences, particularly with deference toward the ID of low plenteous proteins, proteins with very high and low atomic loads, and is somewhat unreasonable regarding enormous scale site explicit examination of post-translational modifications (PTMs), for example, phosphorylation. The innate impediments related to two-dimensional polyacrylamide gel electrophoresis-MS lead to the improvement of without gel MS-based methodologies, otherwise called "shotgun" proteomics. Currently, proteome analysis based on MS is a very powerful and effective tool in the post-genome time. Normal Shotgun proteomic analysis containing four components, preparation of sample, separation, and isolation of peptides, MS analysis of the separated protein fragment, and the analysis of data [26, 27]. Advances made in shotgun based proteomics approaches are generally because of innovative upgrades made in elite mass spectrometers. Since most natural examples comprise of extremely complex peptide blends, mass spectrometers must be fit for guaranteeing a profound logical inclusion while keeping up a significant level of heartiness, affectability, and estimation exactness [28]. More up to date ages of crossbreed mass spectrometers, for example, however not restricted to the LTQ-Orbitrap [29, 30], LTQ Orbitrap Velos [31], Q-Exactive HF [32], and Orbitrap Fusion [33] can accomplish exact mass exactness at high obtaining velocities, and goals which take into account total testing of complex peptide separates giving far-reaching proteome inclusion. Joined with high-performance liquid chromatography (HPLC) innovations, these LC-MS work processes likewise alluded to as sans gel shotgun proteomics takes into consideration the evaluation and ID of whole proteomes just as their protein adjustments crosswise over various organic examples [34, 35]. A typical Shotgun approach has been depicted in Fig. **(3)**.

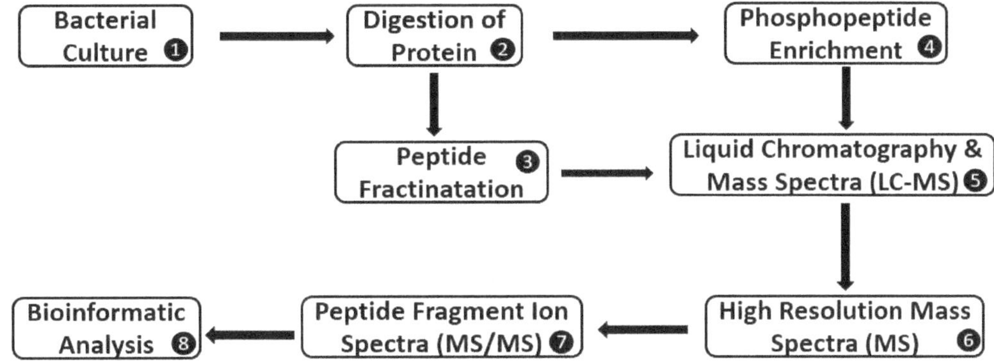

Fig. (3). Shotgun proteomics approach for the analysis of protein. The steps are as follow:
(1) Bacterial cell grown in optimum culture environment and bacterial cells are lysed by different methods, **(2)** The bacterial proteins are digested by proteases enzyme to generate protein fragment (fragmentation), **(3)** The fragmented peptides are fractions to reduce the complexity and to increase the peptide identification and quantification rate, **(4)** In the case of post-translational analysis of peptide, different enrichment process needs to be performed, **(5)** The fractioned peptides are separated through HPLC and high-resolution mass spectrometer (HRMS), **(6)** Comparative quantification of the peptide has to be done, **(7)** Peptides are identified through sequencing and tandem MS, **(8)** Data is processed through different bioinformatic tools, techniques, and approaches.

An ongoing report utilized a combinatorial methodology using 16SrRNA sequencing with ultra-execution fluid chromatography/quadrupole time of flight (UPLC-TOF) MS toward the age of a 3-dimensional geological guide of microorganisms and their atoms of beginning (peptides, metabolites) disseminated on the outside of the human skin [36]. This data was used to distinguish microbial species present and related to the substance condition of the skin and fills in as an incredible methodology toward explaining how the microbiome connects and along these lines changes various regions of the human skin in an animal type's explicit style.

Shotgun MS-Based Quantitative Proteomics

The total identification of proteome and its individual PTMs gives a superior comprehension toward natural capacity and guideline. Be that as it may, it is similarly essential to evaluate elements of proteins comparative with one another under various natural conditions, malady states, and irritations. This data can be acquired using quantitative proteomics of which various methodologies have been built up that are good with shotgun MS-based proteomics techniques. One normal methodology is to present a stable non-radioactive isotope mark either artificially or metabolically on the peptide or protein level. Through this procedure, the general forces of peptides are estimated, employing MS in this way, recognizing and measuring controlled proteins from various examples through the acquired mass spectra. Instances of such systems incorporate yet they are not restricted to Stable Isotope Labeling by Amino acids in Cell culture [37] or 15N marking [38, 39], the two of which have been effectively actualized in an assortment of microscopic organisms [40 - 46]. Moreover, an imaginative MS-based strategy known as cell type-explicit marking utilizing amino corrosive forerunners (CTAP) can constantly name the proteome of individual cell types effectively developing in a co-culture condition, considering the clarification of the "phone of-starting point" of proteins in multicellular situations [47]. Even though this technique was exhibited in eukaryotic models, the application toward microscopic organisms, particularly those in multispecies situations, for example, human microbiomes, biofilms, and so on, could altogether contribute toward the better comprehension of the elements between various bacterial species occupying a similar domain. Albeit metabolic marking approaches, for example, SILAC is considered by numerous individuals to be the most exact strategy toward worldwide protein quantitation [37], they might be hard to execute because of specialized difficulties with the fuse of the compound or metabolic naming methodology, or difficulties presented in complex natural models, for example, those including the skin. In this manner, because of fast advances made in LC-MS, level-free quantitation (LFQ) approaches are presently conceivable and routinely performed [47]. In this methodology, diverse peptide tests are estimated employing MS and analyzed either by the all-out number of sequenced (MS/MS) spectra or the all-out separated particle flows under controlled conditions.

The LFQ procedure was effectively applied to reflect the pathophysiology of *S. aureus* in patients with ectodermal dysplasia and AD [48]. Patients with AD are found to be infected with opportunistic *S. aureus* due to diminished immune system activities and have decreased levels of some of the antimicrobial peptides such as dermcidin leads to decreased immune activities [49]. The creators demonstrated through proteomics examination of the secretome of unhealthy *versus* sound patients that a comparative system of diminished perspiration inferred dermcidin additionally exists in patients with ectodermal dysplasia, making these people profoundly inclined to obtaining *S. aureus* diseases [48].

Quantitative proteomics strategies can be used to evaluate, without a doubt, the molar sum or grouping of a specific protein for every cell. This data is exceptionally significant, particularly inside the clinical setting [50]. Numerous variations of this method exist and all include spiking in a known grouping of an inward protein standard into the example, trailed by MS investigation, and contrasting coming about example peptide estimations with the inner standard. Methods incorporate FLEXIQuant [51], outright evaluation utilizing protein epitope signature labels (PrEST) [52], power-based total measurement (iBAQ) [53], and total measurement (AQUA) [54]. These strategies can be applied to microbes and fill in as a significant apparatus toward understanding the basic instruments during bacterial destructiveness and anti-microbial obstruction.

A few stages are engaged with planning tests for MS, for example, protein extraction, fractionation, processing, partition, and ionization, and each adds to the general variety saw in proteomics information. Furthermore, specialized components like every day and hurry to-run variety in the complex trial gear can make methodical inclinations in the information securing stage.

EXPERIMENTAL PROCEDURE

The enzymatic cleavage of a complex mixture of protein is the first basic step of the LC-MS based proteomics, which is then analyzed through a mass spectrometer. This is the basic difference with the top-down proteomics, which is restricted to the simple protein mixture and does not digest proteins [55]. The steps in the bottom-up proteomics are as follows:

a. Protein extraction from the test sample.
b. Fractionation and Purification of protein to discard contaminants and undesirable and unwanted proteins like housekeeping gene products.
c. Protein digestion through proteases like trypsin which results in small peptides.
d. Separation of the mixture after digestion to get a homogenous mixture of peptides.
e. Mass spectrometer analysis of the peptide mixture.

The two challenges during the mass spectrometer based proteome data analysis are the recognition of the peptide fragment found in the sample mixture and its quantitative estimation. An LC-MS based protein analysis needs some test preparation steps. An LC-MS-based proteomic dissect requires a couple of phases of test course of action, which includes bacterial cell lysis and isolation of components, separation of peptide fragment to spread out the arrangement of proteins into progressively homogenous get-togethers, and protein preparing to break flawless proteins into logically reasonable peptide fragments. Afterwards, when this process is finished, the peptide fragments are ionized and brought into the mass spectrometer for further analysis.

i. Sample Preparation

The analysis of complete cell proteome includes the collection of intact bacterial cells, washing of these cells, the addition of lysate buffer, which contains the plasma membrane lysing molecule and protease inhibitor that stops any kind of proteolysis, to lyse the cells. The lysed cell fragments are homogenized and incubated with the buffer. The incubated mixture is then centrifuged to remove the cell debris and fragments of plasma membrane from the cell lysates. If the test sample is body fluids such as saliva, blood, serum, the cell lysis steps are not required. The blood sample is directly centrifuged to separate blood cells from the plasma and other blood proteins like fibrinogens and clotting factors. High abundant proteins do not play a significant role in the diseases and hence they also need to be removed from the sample; otherwise, it will affect the MS spectra. As the proteome of a cell is very complex, hence the separation step helps in the spreading of protein according to different physical and chemical properties to make it easier to visualize a more number of proteins [56 - 58].

Protein separation and identification are done through gel electrophoresis and facilitated by protein digestion, performed either chemically or enzymatically. This process removes the challenges which come due to the structural complexity of proteins. The smaller peptides are better tractable chemically and due to its smaller size, these peptides are more comfortable to MS analysis. Some of the protein digestive components are trypsin or cyanogen bromide as these digest protein at very specific sites in their chain and hence peptide fragment prediction and theoretical database creation can be possible. The digestion can occur in the gel also, but the preferred medium for digestion is the solution as it is very difficult to separate protein from gel due to their hardness [56]. Any missed cleavage or missed peptide can leads to misidentification of protein when we search against databases.

ii. Mass Spectrometry

The main role of the mass spectrometer is the measurement of mass by charge ration (m/z) for the ionized atoms/molecules. Nowadays, there is tremendous growth in MS technology and almost more than 20 different mass spectrometers are available which can be used for proteomics. All these available spectrometers are developed to carry our different types of ionization and mass spectrometric analysis. The spectrometer consists of three basic units, ion source for the generation and ionization of proteins, mass analyzer for the analysis of protein-based on their m/z ratio and the detector to detect these separated proteins (Fig. **4**). The role of the ion source is to ionize the protein and assign charges on amino acids of protein and the function of the mass analyzer is to analyze different forms of ionized proteins and peptides and measure the mass to charge ratio for every ion. The detector captures each ion and measures its intensity. As a result, we get a mass spectrum which is the representation of mass to charge ratio for each ion captured by mass spectrometer on the x-axis and their peak intensity appears on the y-axis detected by detector [56].

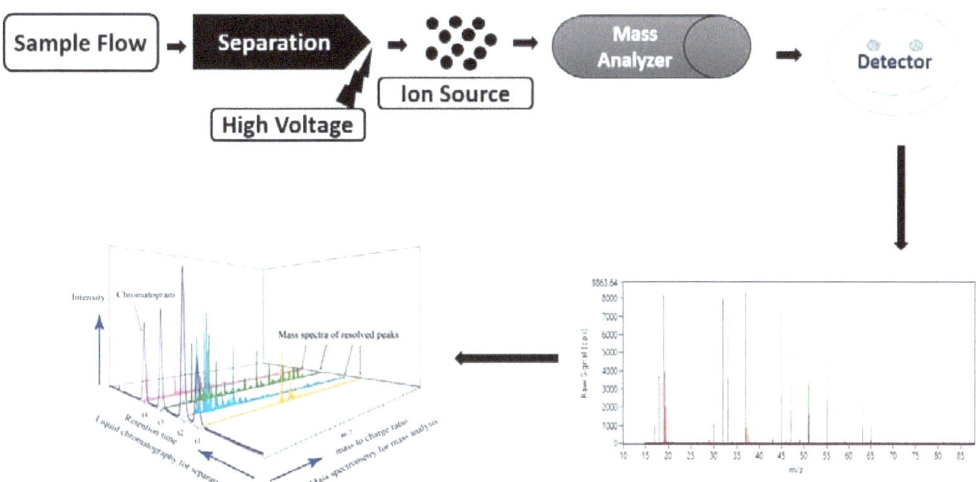

Fig. (4). Schematic diagram of mass spectroscopy. The basic components of the mass spectrometer is an ion source which generates ion and ionizes proteins, a mass analyzer that separates peptide based on mass to charge ration and intensities and a detector that detects these differentiated mass to charge ration for ionized peptide.

There are different ionization techniques used in mass spectrometer some of these are *"electron sway"*, *"substance ionization"*, *"quick iota siege"*, *"field desorption"*, *"electro spray ionization (ESI)"* and *"laser desorption"* and they work by they introduce protons to the peptides. The two most preferred techniques used in the proteome analysis are *"Electro spray ionization"* and *"Matrix-Assisted Laser Dispersion/Ionization (MALDI)"*. In the case of ESI, the liquid sample preparation occurs at atmospheric pressure and they are passed through a very thin needle which is a very high voltage. Electrostatic repulsion causes the dissociation of the solvent drops from the needle tip and sprinkles the charged droplets. After the evaporation of solvent charges, molecules were left alone [56 - 58]. ESI, which is a very delicate ionization technique and it does not break the peptide further, and fragments naturally accommodate the charges peptide in liquid solution; hence it is the most effective interface of LC-MS. In the case of MALDI, the biological molecules are scattered in a crystalline matrix. The crystalline grid is targeted by UV laser beam, which results in the ejection of ionized molecules so that it can be separated into an MS. The mass analyzer is important in terms of the sensitivity (ability to spot the low available protein/peptides), resolution on differentiation (ability to differentiate two or more ion with almost equal mass to charge ratio) and sensitivity MS for mass measurement (ability to measure molecular mass which is very near to the actual value). The most commonly used mass analyzer belongs to *"quadrupole (Q)"*, *"ion-trap (IT)"*, *"time-of-flight (TOF)"*, *"Fourier transform ion cyclotron resonance (FTICR)"*, and *"Orbitrap types"*. Sometimes two mass analyzers can be combined to get bet results such as QTOF, triple-Q [59, 60]. In the case of tandem MA (denoted by MS/MS or MS^n), mass spectroscopy is repeated several times for the same sample. The results coming from this can be used for the identification of proteins. In most of the tandem MS analyzer, the best most abundant peaks can be picked automatically from the original parent MS spectra, and then selected ions will go got further fragmentation and scanning. This step is continued until all the selected peaks from the original scan are done [60 - 62]. The problem of under sampling is not so serious that the use of MS^n for the detection and analysis should be avoided, but we should understand that not all the proteins or peptides are going to be selected for scanning purposes. As the tandem MS is taking much time; hence it is not recommended for the high throughput analysis [63]. The MS/MS-based proteomics is currently one of the very commonly used proteomic techniques and most of the protein detection methodologies are based on this one. Here the tandem MS is run after the protein separation by LC and expressed by LC-MS/MS. *"High-resolution LC-MS instruments (HRMS)(for example, FTICR)"* are exceptionally quick and can accomplish mass estimations that are adequately exact for recognizable proof purposes. As fragmentation and rescanning are not needed; hence the under-sampling issue like tandem MS does not occur in this case. Even though the fragmentation pattern is significant for the identification of peptides, so the hybrid techniques which involve both MS/LC-MS and FTICR are progressively being utilized. One such model is the Accurate Mass and Time (AMT) label approach [64 - 66]. In this case, tandem MS investigation is performed to make an Accurate Mass and Time (AMT) database of peptide hypothetical mass and anticipated elution time, in light of high-certainty IDs from fracture designs, trailed by a solitary mass spectrometer run on FTICR to get exceptionally precise molecular mass estimations, just as LC- elution times; small peptide recognizable proof is then made by contrasting the watched mass estimations and elution times to the AMT database passages. The AMT database is created by the use of several LC-MS/MS runs, which results in a perfect proteome database for proteotypic protein fragments [67]. As AMT uses the database created using previous LC-MS spectra and hence here also the under sampling is not at all an issue.

DATA ACCOMPLISHMENT

In the LC-MS, every sample can result in a large number of output scans, which contains a mass spectrum. The single MS scan mass spectrum is then depicted by the plot of mass to charge ratio (m/z) on the x-axis and peak intensities of separated masses on the y-axis. These data contain the signal generated from each peptide. As a first step to recognize and evaluate those peptides, highlights should be distinguished in the information and, for instance, recognized from background non-important peaks. One of the first and important steps in this is the detection of MS peak in the MS spectrum. There are various methods have been suggested for peak detection. Some of the labs use the signal to noise ratio for a peak compared to its local backgrounds. Every fragmented peptide gives an envelope of peaks as peptides are composed of amino acids. The occurrence of peptide peak can be found by the mass-to-charge ratio against the peak appears from the most prevalent isotopes, which are called monoisotopic mass. There may be many isotopes present for an element that makes amino acids; ^{13}C is more prevalent among all the isotopes. ^{12}C and ^{13}C have a difference in their mass by 1 Dalton, the gap between the monoisotopic peak of peptide will be 1/z from the ^{13}C peak.

The deisotoping of a mass spectrum is a process to remove the redundant information from the isotopic MS peaks and to simplify the data. The following steps are to be taken care of for the deisotoping process:

i. Finding the isotopic dispersion in a mass spectrum.
ii. Processing the charge condition of every peptide dependent on the separation between peaks in their isotopic dispersion, and
iii. Monoisotopic mass of every peptide fragments should be taken into consideration .

One important point needs to be noted that this process can be possible only if the HRMS measurement has been taken otherwise, the isotopic data and corresponding peaks cannot be disclosed. For the finding of isotopic distribution on spectra, the identified peaks are considered as the potential peak from isotopic dissemination, and hypothetical isotropic dispersions got from a database of peptide groupings, are overlaid with the watched spectra. One major challenge in deisotoping is the occurrence of overlapping isotopic distribution from various peptides. There are several methodologies and algorithm is available for spotting the peak and deionizing the isotopic peaks. Some commercially available tools are provided by vendors like Agilent, Thermo Fisher, Rosetta, Decon2LS, *etc.* [68].

A peptide will wash out from the HPLC in several scans, which create an elution profile. The elution profile of a peptide is normally short in time-lapse and functions to define a characterize a significant component in LC-MS data sets. There might be contaminants also present in the sample which might have a very long elution profile and it can be washed out in preprocessing steps. There are various approaches available, which deal with outlining an elution profile, such as NET Elution Time (NET) [69]. At this point, every LC-MS test sample is with a relegated monoisotopic mass and elution time. Sometimes due to error in the mass measurement and elution time, the assigned pair (mass and elution time) may vary between different samples. Alignment is frequently performed to arrange the LC-MS includes in various examples such as Crawdad [70] and LCMSWarp [71]. Along with all the high throughput analysis- omics techniques, mass spectrometer based proteomic information is ordinarily exposed to significant preprocessing and standardization. Deliberate predispositions are frequently found in mass estimations, elution times, and pinnacle powers [72, 73]. Permeating of low standard proteins is also normal [74]. In standardization, care must be taken to isolate organic sign from specialized predisposition [75]. Broadly utilized standardization systems in high-throughput genomic or proteomic think about include some variety of worldwide scaling, scatter plot smoothing, or ANOVA [76]. Worldwide scaling, for the most part, includes moving every one of the estimations for a solitary example by a steady sum, so the methods, medians, or complete particle flows (TICs) of all examples are equal. Since normally specialized predispositions are more perplexing than basic moves between tests, worldwide scaling cannot catch complex inclination highlights. Scatter plot smoothing, TIC, and ANOVA standardization strategies are tested explicitly and consequently, progressively adaptable. An as of late proposed strategy, called EigenMS, evacuates inclination of subjective multifaceted nature by the utilization of the solitary worth deterioration to catch and expel predispositions from LC-MS top power estimations [77]. EigenMS expels predispositions of discretionary intricacy and modifies the standardized powers to address the p-values after standardization (guaranteeing that invalid p-values are consistently disseminated).

The manufacturers of Mass spectrometers have designed and developed several binary data formats for the storage of the output file and data from the instruments, such as *".baf (Bruker)", ".Raw (Thermo)"* and *".PKM (Applied Biosystems)".* Dealing data from different binary data format needs a corresponding proprietary software making it hard to share datasets. These limitations have been addressed by the development of some XML-based seller free information groups such as seller free information groups as mzXML [78, 79], mz Data [80], and mzML [81, 82]. mzML 1.0 is viewed as a converging of the best of mzData and mzXML and this was released in June 2009. This file output format keeps the information about spectra, instruments used for analysis, setting of the instruments Dents during analysis, and also the data processing details.

PROTEIN IDENTIFICATION

In the case of bottom-up proteomics, the peptide can be detected by the comparison of noted spectroscopy data to the protein MS database which contains data from the predicted or previously identified protein. The preferred techniques are tandem MS along with the database looking [83], in which protein dissociation patterns are contrasted with hypothetical examples in a database utilizing programming like *"Sequest"* [84], *"X!Tandem"* [85], and *"Mascot"* [86]. With HR LC-MS analyzers, the identification of the peptide is performed with only mass and elution time alone or in combination with tandem-MS technology. The other option to find the database includes peptide sequencing by *de novo* methodology [87, 88] and the combination of both *de novo* and database searching approach [89, 90]. In the case of tandem MS, the precursor ion for the most prevalent peaks in a scan is fragmented and rescanned. In the case of collision-induced dissociation (CID), the precursor ions are dissociated into fragment by collision with a neutral gas [91 - 93]. Then the MS analysis measures the m/z ration on the x-axis and intensity on the y-axis of the fragmented fragments, leads to the generation of a fragment pattern. Generally, CID generates the y-ion and b-ion by the breakage of amide bonds in a peptide backbone. When the charge is kept by the amino-terminal fragment, it forms b-ion and when the charge is kept by the carboxy-terminal fragment, it forms y-ion. Fragmentation nearby some specific amino acid is very common and these amino acids are glutamic acid (E), aspartic acid (D), and proline (P). Some other types of fragmentation patterns are also possible such as a-, c-,x-,, and z-types. Electron capture dissociation (ECD) generates c-ion and z-ions and keeps the side chain of amino acids intact. The pattern on which the peptide is going to be fragmented is the fingerprint of the peptide chain. As, it is represented by the sequence of amino acids and hence fragmentation pattern can be predicted. The detected fragmented pattern should match with the theoretically assumed fragments.

The right match can only be made if the correct sequence is in the database in any case. If the genome of an organism is not complete or having some kind of error, this will not be the situation. Furthermore, due to the under-sampling issue in case of tandem MS, only a very less percentage of peptide available in the sample will be considered for detection purposes. The main reason for this is that only a very few portions of the available peaks are going to be selected from the spectra in the first mass spectroscopy step for the fragmentations in the second mass spectroscopy studies. Along these lines, low occurred peaks are neglected by the presence of high prevalent peaks. *De novo* sequencing, an alternate approach for database search, is the assembly of amino acids of the peptide-based on direct inspection of mass spectra [87, 88]. For a given sequence of amino acid, the possible atomization ions and masses can be enumerated and also the awaited frequency with which each fragment formed. The major difference with the database search is that we do not need prior sequence information. The drawback associated with the *de novo* approach is the high cost for the computational system and also the requirement of more sample quantity.

USE OF MALDI-TOF IN MICROBIAL IDENTIFICATION AND ANTIMICROBIAL RESISTANCE

MALDI-TOF mass spectrometer is the most preferred tool for the identification of bacterial pathogenic strain from the blood culture due to its fast and rapid performance and a very small amount of microbial mass requirements [94 - 96]. To perform this, the separation of pathogens from the bacterial culture is one of the critical steps. Several protocols for the sample preparation steps have been proposed, one such technique is through cell lysis or differential centrifugation. Alternate to the lytic method is the purification of bacteria from the blood culture with the uses of serum separator tubes. Compared to cell lysis protocol, this is the simple, fast, and high yielding process [96]. There are several studies that have been done where, along with MALDI-TOF MS and microbial stewardship intervention for the contouring empirical antibacterial treatment in the patients with a blood infection, can improve patients' outcomes significantly [97 - 100]. A rapid MALDI-TOF MS identification can be used in association with the fast molecular method for the identification and evaluation of antimicrobial resistance [95, 96]. The LCMS/MS-based proteomics and bioinformatics tools have also been used for the analysis of complete proteomics in carbapenem-resistant *K. Pneumoniae* under stress created by the use of meropenem [101, 102]. The beta-lactam ring hydrolysis can find out by MS with the absence of an original peak in mass spectra of the antibiotic and appearance of the subsequent hydrolytic product [103].

CONCLUSION

Microbial proteomics is going to be one of the compulsory components in the translational research, developing clinical diagnostic tools, antibacterial treatment strategies, and vaccine development. Along with these applications, microbial proteomics is also providing support in the elucidation of bacterial metabolic pathways and the relationship between the microorganisms and human health. The microbial proteomics has seen tremendous growth these days. In the last few decades, microbial proteomics is worked very much and these are focussed towards fundamental research to applied research, pathological research to functional research. There are several tools and instrumentation is available for the study and analysis of microbial proteomics such as conventional chromatography to mass spectroscopy and protein microarray-based proteomic analysis. These techniques opened a new horizon of information on the expression of the protein in the nucleus, its occurrence, its post-translational modifications, localization, and also various types of interactions such as protein-protein, protein-DNA, enzyme-substrate, substrate-ligands, *etc.* The ability of microbes to bear different types of stresses, which might be environmental, physiological, toxic compounds, pollutants, host immune systems, or immunological components, is beneficial for fundamental and applied research. The proteomics technologies provide information on the understanding of the molecular mechanism of these stress-bearing mechanisms. Although the LC-MS based proteomics is developing very fast these days, there are still a lot of challenges in the proteome analysis. The complication of proteome and the countless bioinformatic tasks need to be performed to convert the test sample into analyzable data and increase the reproducibility. This information can be used for drug development or vaccine development purposes. The combination of genomics and proteomics data helps to improve the annotation of the genome by supplying information about alternate splicing and post-translational modifications. However, much more work needs to be done to improve the reproducibility and performance of well-known proteomics tools.

CONSENT FOR PUBLICATION

Not applicable.

CONFLICT OF INTEREST

The author declares no conflict of interest, financial or otherwise.

ACKNOWLEDGEMENTS

Declared none.

REFERENCES

[1]   Han MJ, Lee JW, Lee SY. Understanding and engineering of microbial cells based on proteomics and its conjunction with other omics studies. Proteomics 2011; 11(4): 721-43.
[http://dx.doi.org/10.1002/pmic.201000411] [PMID: 21229587]

[2]   Cristea IM, Gaskell SJ, Whetton AD. Proteomics techniques and their application to hematology. Blood 2004; 103(10): 3624-34.
[http://dx.doi.org/10.1182/blood-2003-09-3295] [PMID: 14726377]

[3]   Wilkins MR, Sanchez JC, Gooley AA, *et al.* Progress with proteome projects: why all proteins expressed by a genome should be identified and how to do it. Biotechnol Genet Eng Rev 1996; 13(1): 19-50.
[http://dx.doi.org/10.1080/02648725.1996.10647923] [PMID: 8948108]

[4]   Pandey A, Mann M. Proteomics to study genes and genomes. Nature 2000; 405(6788): 837-46.
[http://dx.doi.org/10.1038/35015709] [PMID: 10866210]

[5]   Domon B, Aebersold R. Mass spectrometry and protein analysis. Science 2006; 312(5771): 212-7.
[http://dx.doi.org/10.1126/science.1124619] [PMID: 16614208]

[6]   Krishna RG, Wold F. Post-translational modification of proteins. Advances in Enzymology and Related Areas of Molecular Biology. Hoboken, NJ, USA: Wiley-Blackwell 1993; pp. 265-98.

[7]   Lander ES, Linton LM, Birren B, *et al.* Erratum: Initial sequencing and analysis of the human genome: International Human Genome Sequencing Consortium (Nature (2001) 409 (860-921)). Nature 2001; 412(6846): 565-6.

[8]   Canales RD, Luo Y, Willey JC, *et al.* Evaluation of DNA microarray results with quantitative gene expression platforms. Nat Biotechnol 2006; 24(9): 1115-22.
[http://dx.doi.org/10.1038/nbt1236] [PMID: 16964225]

[9]   Jungbauer A, Hahn R. Ion-exchange chromatography InMethods in enzymology. Academic Press 2009; Vol. 463: pp. 349-71.

[10]   Ong SE, Mann M. Mass spectrometry-based proteomics turns quantitative. Nat Chem Biol 2005; 1(5): 252-62.
[http://dx.doi.org/10.1038/nchembio736] [PMID: 16408053]

[11]   Morris JS, Clark BN, Gutstein HB. Pinnacle: a fast, automatic and accurate method for detecting and quantifying protein spots in 2-dimensional gel electrophoresis data. Bioinformatics 2008; 24(4): 529-36.
[http://dx.doi.org/10.1093/bioinformatics/btm590] [PMID: 18194961]

[12]   Nesvizhskii AI, Vitek O, Aebersold R. Analysis and validation of proteomic data generated by tandem mass spectrometry. Nat Methods 2007; 4(10): 787-97.
[http://dx.doi.org/10.1038/nmeth1088] [PMID: 17901868]

[13]   Tang N, Tornatore P, Weinberger SR. Current developments in SELDI affinity technology. Mass Spectrom Rev 2004; 23(1): 34-44.
[http://dx.doi.org/10.1002/mas.10066] [PMID: 14625891]

[14]   Kašička V. Recent developments in capillary and microchip electroseparations of peptides (2017-mid 2019). Electrophoresis 2020; 41(1-2): 10-35.
[http://dx.doi.org/10.1002/elps.201900269] [PMID: 31657477]

[15]   Perry WJ, Weiss A, Van de Plas R, Spraggins JM, Caprioli RM, Skaar EP. Integrated molecular imaging technologies for investigation of metals in biological systems: A brief review. Curr Opin Chem Biol 2020; 55: 127-35.
[http://dx.doi.org/10.1016/j.cbpa.2020.01.008] [PMID: 32087551]

[16]   Stoeckli M, Chaurand P, Hallahan DE, Caprioli RM. Imaging mass spectrometry: a new technology for the analysis of protein expression in mammalian tissues. Nat Med 2001; 7(4): 493-6.
[http://dx.doi.org/10.1038/86573] [PMID: 11283679]

[17]   Jackson SN, Muller L, Roux A, *et al.* AP-MALDI mass spectrometry imaging of gangliosides using 2, 6-dihydroxyacetophenone. J Am Soc Mass Spectrom 2018; 29(7): 1463-72.
[http://dx.doi.org/10.1007/s13361-018-1928-8] [PMID: 29549666]

[18]   Griffiths WJ, Jonsson AP, Liu S, Rai DK, Wang Y. Electrospray and tandem mass spectrometry in biochemistry. Biochem J 2001; 355(Pt 3): 545-61.
[http://dx.doi.org/10.1042/bj3550545] [PMID: 11311115]

[19]   Poland GA, Ovsyannikova IG, Johnson KL, Naylor S. The role of mass spectrometry in vaccine development. Vaccine 2001; 19(17-19): 2692-700.
[http://dx.doi.org/10.1016/S0264-410X(00)00505-3] [PMID: 11257411]

[20]   Fenn JB, Mann M, Meng CK, Wong SF, Whitehouse CM. Electrospray ionization for mass spectrometry of large biomolecules. Science 1989; 246(4926): 64-71.
[http://dx.doi.org/10.1126/science.2675315] [PMID: 2675315]

[21]   Mann M. The ever expanding scope of electrospray mass spectrometry-a 30 year journey. Nat Commun 2019; 10(1): 3744.
[http://dx.doi.org/10.1038/s41467-019-11747-z] [PMID: 31501421]

[22]   Yates III Jr. A century of mass spectrometry: from atoms to proteomes. Nat Methods 2011; 8(8): 633-7.
[http://dx.doi.org/10.1038/nmeth.1659]

[23]   Biswas S, Rolain JM. Use of MALDI-TOF mass spectrometry for identification of bacteria that are difficult to culture. J Microbiol Methods 2013; 92(1): 14-24.
[http://dx.doi.org/10.1016/j.mimet.2012.10.014] [PMID: 23154044]

[24]   Trauger SA, Junker T, Siuzdak G. Investigating viral proteins and intact viruses with mass spectrometry InModern Mass Spectrometry. Berlin, Heidelberg: Springer 2003; pp. 265-82.

[25]   Ribeiro FJ, Przybylski D, Yin S, *et al.* Finished bacterial genomes from shotgun sequence data. Genome Res 2012; 22(11): 2270-7.
[http://dx.doi.org/10.1101/gr.141515.112] [PMID: 22829535]

[26]   Ye X, Tang J, Mao Y, *et al.* Integrated Proteomics Sample Preparation and Fractionation: Method Development and Applications. Trends Analyt Chem 2019; 115667.
[http://dx.doi.org/10.1016/j.trac.2019.115667]

[27]   Ahmed F, Kumar G, Soliman FM, *et al.* Proteomics for understanding pathogenesis, immune modulation and host pathogen interactions in aquaculture. Comp Biochem Phys D: Genom Proteom 2019; 100625
[http://dx.doi.org/10.1016/j.cbd.2019.100625]

[28]   Mann M, Kelleher NL. Precision proteomics: the case for high resolution and high mass accuracy. Proc Natl Acad Sci USA 2008; 105(47): 18132-8.
[http://dx.doi.org/10.1073/pnas.0800788105] [PMID: 18818311]

[29]   Mohler RE, Ahn S, O'Reilly K, *et al.* Towards comprehensive analysis of oxygen containing organic compounds in groundwater at a crude oil spill site using GC×GC-TOFMS and Orbitrap ESI-MS. Chemosphere 2020; 244: 125504.
[http://dx.doi.org/10.1016/j.chemosphere.2019.125504] [PMID: 31837566]

[30]   Hu Q, Noll RJ, Li H, Makarov A, Hardman M, Graham Cooks R. The Orbitrap: a new mass spectrometer. J Mass Spectrom 2005; 40(4): 430-43.
[http://dx.doi.org/10.1002/jms.856] [PMID: 15838939]

[31]   Olsen JV, Schwartz JC, Griep-Raming J, *et al.* A dual pressure linear ion trap Orbitrap instrument with very high sequencing speed. Mol Cell Proteomics 2009; 8(12): 2759-69.
[http://dx.doi.org/10.1074/mcp.M900375-MCP200] [PMID: 19828875]

[32]   Passarelli MK, Pirkl A, Moellers R, *et al.* The 3D OrbiSIMS-label-free metabolic imaging with subcellular lateral resolution and high mass-resolving power. Nat Methods 2017; 14(12): 1175-83.
[http://dx.doi.org/10.1038/nmeth.4504] [PMID: 29131162]

[33]   Erickson BK, Jedrychowski MP, McAlister GC, Everley RA, Kunz R, Gygi SP. Evaluating multiplexed quantitative phosphopeptide analysis on a hybrid quadrupole mass filter/linear ion trap/orbitrap mass spectrometer. Anal Chem 2015; 87(2): 1241-9.
[http://dx.doi.org/10.1021/ac503934f] [PMID: 25521595]

[34]   Bek-Thomsen M, Lomholt HB, Scavenius C, Enghild JJ, Brüggemann H. Proteome analysis of human sebaceous follicle infundibula extracted from healthy and acne-affected skin. PLoS One 2014; 9(9): e107908.
[http://dx.doi.org/10.1371/journal.pone.0107908] [PMID: 25238151]

[35]   Williams HC, Dellavalle RP, Garner S. Acne vulgaris. Lancet 2012; 379(9813): 361-72.
[http://dx.doi.org/10.1016/S0140-6736(11)60321-8] [PMID: 21880356]

[36]   Bouslimani A, Porto C, Rath CM, *et al.* Molecular cartography of the human skin surface in 3D. Proc Natl Acad Sci USA 2015; 112(17): E2120-9.
[http://dx.doi.org/10.1073/pnas.1424409112] [PMID: 25825778]

[37]   Ong SE, Blagoev B, Kratchmarova I, *et al.* Stable isotope labeling by amino acids in cell culture, SILAC, as a simple and accurate approach to expression proteomics. Mol Cell Proteomics 2002; 1(5): 376-86.
[http://dx.doi.org/10.1074/mcp.M200025-MCP200] [PMID: 12118079]

[38]   Hempel K, Pané-Farré J, Otto A, Sievers S, Hecker M, Becher D. Quantitative cell surface proteome profiling for SigB-dependent protein expression in the human pathogen *Staphylococcus aureus via* biotinylation approach. J Proteome Res 2010; 9(3): 1579-90.
[http://dx.doi.org/10.1021/pr901143a] [PMID: 20108986]

[39]   Soufi B, Krug K, Harst A, Macek B. Characterization of the *E. coli* proteome and its modifications during growth and ethanol stress. Front Microbiol 2015; 6: 103.
[http://dx.doi.org/10.3389/fmicb.2015.00103] [PMID: 25741329]

[40]   Soufi B, Kumar C, Gnad F, Mann M, Mijakovic I, Macek B. Stable isotope labeling by amino acids in cell culture (SILAC) applied to quantitative proteomics of *Bacillus subtilis*. J Proteome Res 2010; 9(7): 3638-46.
[http://dx.doi.org/10.1021/pr100150w] [PMID: 20509597]

[41]   Hempel K, Herbst FA, Moche M, Hecker M, Becher D. Quantitative proteomic view on secreted, cell surface-associated, and cytoplasmic proteins of the methicillin-resistant human pathogen *Staphylococcus aureus* under iron-limited conditions. J Proteome Res 2011; 10(4): 1657-66.
[http://dx.doi.org/10.1021/pr1009838] [PMID: 21323324]

[42]   Soares NC, Spät P, Krug K, Macek B. Global dynamics of the *Escherichia coli* proteome and phosphoproteome during growth in minimal medium. J Proteome Res 2013; 12(6): 2611-21.
[http://dx.doi.org/10.1021/pr3011843] [PMID: 23590516]

[43]   Misra SK, Moussan Désirée Aké F, Wu Z, *et al.* Quantitative proteome analyses identify PrfA-responsive proteins and phosphoproteins in Listeria monocytogenes. J Proteome Res 2014; 13(12): 6046-57.
[http://dx.doi.org/10.1021/pr500929u] [PMID: 25383790]

[44]   Soufi B, Macek B. Stable isotope labeling by amino acids applied to bacterial cell culture InStable Isotope Labeling by Amino Acids in Cell Culture (SILAC). New York, NY: Humana Press 2014; pp. 9-22.

[45]   Boysen A, Borch J, Krogh TJ, Hjernø K, Møller-Jensen J. SILAC-based comparative analysis of pathogenic *Escherichia coli* secretomes. J Microbiol Methods 2015; 116: 66-79.
[http://dx.doi.org/10.1016/j.mimet.2015.06.015] [PMID: 26143086]

[46]   Gauthier NP, Soufi B, Walkowicz WE, *et al.* Cell-selective labeling using amino acid precursors for proteomic studies of multicellular environments. Nat Methods 2013; 10(8): 768-73.
[http://dx.doi.org/10.1038/nmeth.2529] [PMID: 23817070]

[47]   Neilson KA, Ali NA, Muralidharan S, *et al.* Less label, more free: approaches in label-free quantitative mass spectrometry. Proteomics 2011; 11(4): 535-53.
[http://dx.doi.org/10.1002/pmic.201000553] [PMID: 21243637]

[48]   Burian M, Velic A, Matic K, *et al.* Quantitative proteomics of the human skin secretome reveal a reduction in immune defense mediators in ectodermal dysplasia patients. J Invest Dermatol 2015; 135(3): 759-67.
[http://dx.doi.org/10.1038/jid.2014.462] [PMID: 25347115]

[49]   Schittek B. The antimicrobial skin barrier in patients with atopic dermatitis. Curr Probl Dermatol 2011; 41: 54-67.
[http://dx.doi.org/10.1159/000323296] [PMID: 21576947]

[50]   Rodríguez-Suárez E, Whetton AD. The application of quantification techniques in proteomics for biomedical research. Mass Spectrom Rev 2013; 32(1): 1-26.
[http://dx.doi.org/10.1002/mas.21347] [PMID: 22847841]

[51]   Singh S, Springer M, Steen J, Kirschner MW, Steen H. FLEXIQuant: a novel tool for the absolute quantification of proteins, and the simultaneous identification and quantification of potentially modified peptides. J Proteome Res 2009; 8(5): 2201-10.
[http://dx.doi.org/10.1021/pr800654s] [PMID: 19344176]

[52]   Edfors F, Boström T, Forsström B, *et al.* Immunoproteomics using polyclonal antibodies and stable isotope-labeled affinity-purified recombinant proteins. Mol Cell Proteomics 2014; 13(6): 1611-24.
[http://dx.doi.org/10.1074/mcp.M113.034140] [PMID: 24722731]

[53]   Schwanhäusser B, Busse D, Li N, *et al.* Global quantification of mammalian gene expression control. Nature 2011; 473(7347): 337-42.
[http://dx.doi.org/10.1038/nature10098] [PMID: 21593866]

[54]   Gerber SA, Rush J, Stemman O, Kirschner MW, Gygi SP. Absolute quantification of proteins and phosphoproteins from cell lysates by tandem MS. Proc Natl Acad Sci USA 2003; 100(12): 6940-5.
[http://dx.doi.org/10.1073/pnas.0832254100] [PMID: 12771378]

[55]   Han X, Aslanian A, Yates JR III. Mass spectrometry for proteomics. Curr Opin Chem Biol 2008; 12(5): 483-90.
[http://dx.doi.org/10.1016/j.cbpa.2008.07.024] [PMID: 18718552]

[56]   Karpievitch YV, Polpitiya AD, Anderson GA, Smith RD, Dabney AR. Liquid chromatography mass spectrometry-based proteomics: biological and technological aspects. Ann Appl Stat 2010; 4(4): 1797-823.
[http://dx.doi.org/10.1214/10-AOAS341] [PMID: 21593992]

[57]   Berth M, Moser FM, Kolbe M, Bernhardt J. The state of the art in the analysis of two-dimensional gel electrophoresis images. Appl Microbiol Biotechnol 2007; 76(6): 1223-43.
[http://dx.doi.org/10.1007/s00253-007-1128-0] [PMID: 17713763]

[58]   Görg A, Weiss W, Dunn MJ. Current two-dimensional electrophoresis technology for proteomics. Proteomics 2004; 4(12): 3665-85.
[http://dx.doi.org/10.1002/pmic.200401031] [PMID: 15543535]

[59]   Siuzdak G. The Expanding Role of Mass Spectrometry in Biotechnology. San Diego: MCC Press 2003.

[60]   Sharma D, Bisht D. Secretory proteome analysis of streptomycin-resistant *Mycobacterium tuberculosis* clinical isolates. SLAS DISCOVERY: Advancing Life Sciences R&D 2017 Dec; 22; 22(10): 1229-38.
[http://dx.doi.org/10.1177/2472555217698428]

[61]   Zhang H, Yi EC, Li XJ, *et al.* High throughput quantitative analysis of serum proteins using glycopeptide capture and liquid chromatography mass spectrometry. Mol Cell Proteomics 2005; 4(2): 144-55.
[http://dx.doi.org/10.1074/mcp.M400090-MCP200] [PMID: 15608340]

[62]   Garza S, Moini M. Analysis of complex protein mixtures with improved sequence coverage using (CE-MS/MS)n. Anal Chem 2006; 78(20): 7309-16.
[http://dx.doi.org/10.1021/ac0612269] [PMID: 17037937]

[63]   Masselon CD, Kieffer-Jaquinod S, Brugière S, Dupierris V, Garin J. Influence of mass resolution on species matching in accurate mass and retention time (AMT) tag proteomics experiments. Rapid Commun Mass Spectrom 2008; 22(7): 986-92.
[http://dx.doi.org/10.1002/rcm.3447] [PMID: 18320544]

[64]   Pasa-Tolić L, Masselon C, Barry RC, Shen Y, Smith RD. Proteomic analyses using an accurate mass and time tag strategy. Biotechniques 2004; 37(4): 621-624, 626-633, 636 passim.

[http://dx.doi.org/10.2144/04374RV01] [PMID: 15517975]

[65]   Yanofsky CM, Kearney RE, Lesimple S, *et al.* A Bayesian approach to peptide identification using accurate mass and time tags from LC-FTICR-MS proteomics experiments. Annu Int Conf IEEE Eng Med Biol Soc 2008; 2008: 3775-8.
[http://dx.doi.org/10.1109/IEMBS.2008.4650030] [PMID: 19163533]

[66]   Tolmachev AV, Monroe ME, Purvine SO, *et al.* Characterization of strategies for obtaining confident identifications in bottom-up proteomics measurements using hybrid FTMS instruments. Anal Chem 2008; 80(22): 8514-25.
[http://dx.doi.org/10.1021/ac801376g] [PMID: 18855412]

[67]   Mallick P, Schirle M, Chen SS, *et al.* Computational prediction of proteotypic peptides for quantitative proteomics. Nat Biotechnol 2007; 25(1): 125-31.
[http://dx.doi.org/10.1038/nbt1275] [PMID: 17195840]

[68]   Jaitly N, Mayampurath A, Littlefield K, Adkins JN, Anderson GA, Smith RD. Decon2LS: An open-source software package for automated processing and visualization of high resolution mass spectrometry data. BMC Bioinformatics 2009; 10(1): 87.
[http://dx.doi.org/10.1186/1471-2105-10-87] [PMID: 19292916]

[69]   Petritis K, Kangas LJ, Yan B, *et al.* Improved peptide elution time prediction for reversed-phase liquid chromatography-MS by incorporating peptide sequence information. Anal Chem 2006; 78(14): 5026-39.
[http://dx.doi.org/10.1021/ac060143p] [PMID: 16841926]

[70]   Finney GL, Blackler AR, Hoopmann MR, Canterbury JD, Wu CC, MacCoss MJ. Label-free comparative analysis of proteomics mixtures using chromatographic alignment of high-resolution muLC-MS data. Anal Chem 2008; 80(4): 961-71.
[http://dx.doi.org/10.1021/ac701649e] [PMID: 18189369]

[71]   Jaitly N, Monroe ME, Petyuk VA, Clauss TR, Adkins JN, Smith RD. Robust algorithm for alignment of liquid chromatography-mass spectrometry analyses in an accurate mass and time tag data analysis pipeline. Anal Chem 2006; 78(21): 7397-409.
[http://dx.doi.org/10.1021/ac052197p] [PMID: 17073405]

[72]   Callister SJ, Barry RC, Adkins JN, *et al.* Normalization approaches for removing systematic biases associated with mass spectrometry and label-free proteomics. J Proteome Res 2006; 5(2): 277-86.
[http://dx.doi.org/10.1021/pr0503001] [PMID: 16457593]

[73]   Petyuk VA, Jaitly N, Moore RJ, *et al.* Elimination of systematic mass measurement errors in liquid chromatography-mass spectrometry based proteomics using regression models and a priori partial knowledge of the sample content. Anal Chem 2008; 80(3): 693-706.
[http://dx.doi.org/10.1021/ac701863d] [PMID: 18163597]

[74]   Karpievitch Y, Stanley J, Taverner T, *et al.* A statistical framework for protein quantitation in bottom-up MS-based proteomics. Bioinformatics 2009; 25(16): 2028-34.
[http://dx.doi.org/10.1093/bioinformatics/btp362] [PMID: 19535538]

[75]   Dabney AR, Storey JD. A reanalysis of a published Affymetrix GeneChip control dataset. Genome Biol 2006; 7(3): 401.
[http://dx.doi.org/10.1186/gb-2006-7-3-401] [PMID: 16563185]

[76]   Quackenbush J. Microarray data normalization and transformation. Nat Genet 2002; 32(4) (Suppl.): 496-501.
[http://dx.doi.org/10.1038/ng1032] [PMID: 12454644]

[77]   Karpievitch YV, Taverner T, Adkins JN, *et al.* Normalization of peak intensities in bottom-up MS-based proteomics using singular value decomposition. Bioinformatics 2009; 25(19): 2573-80.
[http://dx.doi.org/10.1093/bioinformatics/btp426] [PMID: 19602524]

[78]   Lin SM, Zhu L, Winter AQ, Sasinowski M, Kibbe WA. What is mzXML good for? Expert Rev

Proteomics 2005; 2(6): 839-45.
[http://dx.doi.org/10.1586/14789450.2.6.839] [PMID: 16307524]

[79] Pedrioli PG, Eng JK, Hubley R, *et al.* A common open representation of mass spectrometry data and its application to proteomics research. Nat Biotechnol 2004; 22(11): 1459-66.
[http://dx.doi.org/10.1038/nbt1031] [PMID: 15529173]

[80] Jiménez RC, Vizcaíno JA. Proteomics data exchange and storage: the need for common standards and public repositories InMass Spectrometry Data Analysis in Proteomics. Totowa, NJ: Humana Press 2013; pp. 317-33.

[81] Orchard S, Hoogland C, Bairoch A, Eisenacher M, Kraus HJ, Binz PA. Managing the data explosion. A report on the HUPO-PSI Workshop. August 2008, Amsterdam, The Netherlands. Proteomics 2009; 9(3): 499-501.
[http://dx.doi.org/10.1002/pmic.200800838] [PMID: 19132688]

[82] Deutsch E. mzML: a single, unifying data format for mass spectrometer output. Proteomics 2008; 8(14): 2776-7.
[http://dx.doi.org/10.1002/pmic.200890049] [PMID: 18655045]

[83] Deutsch EW, Lam H, Aebersold R. Data analysis and bioinformatics tools for tandem mass spectrometry in proteomics. Physiol Genomics 2008; 33(1): 18-25.
[http://dx.doi.org/10.1152/physiolgenomics.00298.2007] [PMID: 18212004]

[84] Eng JK, McCormack AL, Yates JR. An approach to correlate tandem mass spectral data of peptides with amino acid sequences in a protein database. J Am Soc Mass Spectrom 1994; 5(11): 976-89.
[http://dx.doi.org/10.1016/1044-0305(94)80016-2] [PMID: 24226387]

[85] Craig R, Beavis RC. TANDEM: matching proteins with tandem mass spectra. Bioinformatics 2004; 20(9): 1466-7.
[http://dx.doi.org/10.1093/bioinformatics/bth092] [PMID: 14976030]

[86] Perkins DN, Pappin DJ, Creasy DM, Cottrell JS. Probability-based protein identification by searching sequence databases using mass spectrometry data. Electrophoresis 1999; 20(18): 3551-67.
[http://dx.doi.org/10.1002/(SICI)1522-2683(19991201)20:18<3551::AID-ELPS3551>3.0.CO;2-2] [PMID: 10612281]

[87] Dancík V, Addona TA, Clauser KR, Vath JE, Pevzner PA. *De novo* peptide sequencing *via* tandem mass spectrometry. J Comput Biol 1999; 6(3-4): 327-42.
[http://dx.doi.org/10.1089/106652799318300] [PMID: 10582570]

[88] Lu B, Chen T. A suboptimal algorithm for *de novo* peptide sequencing *via* tandem mass spectrometry. J Comput Biol 2003; 10(1): 1-12.
[http://dx.doi.org/10.1089/106652703763255633] [PMID: 12676047]

[89] Tanner S, Shu H, Frank A, *et al.* InsPecT: identification of posttranslationally modified peptides from tandem mass spectra. Anal Chem 2005; 77(14): 4626-39.
[http://dx.doi.org/10.1021/ac050102d] [PMID: 16013882]

[90] Frank A, Pevzner P. PepNovo: *de novo* peptide sequencing *via* probabilistic network modeling. Anal Chem 2005; 77(4): 964-73.
[http://dx.doi.org/10.1021/ac048788h] [PMID: 15858974]

[91] Wells JM, McLuckey SA. Collision-induced dissociation (CID) of peptides and proteins. Methods Enzymol 2005; 402: 148-85.
[http://dx.doi.org/10.1016/S0076-6879(05)02005-7] [PMID: 16401509]

[92] Laskin J, Futrell JH. Collisional activation of peptide ions in FT-ICR mass spectrometry. Mass Spectrom Rev 2003; 22(3): 158-81.
[http://dx.doi.org/10.1002/mas.10041] [PMID: 12838543]

[93] Sleno L, Volmer DA. Ion activation methods for tandem mass spectrometry. J Mass Spectrom 2004; 39(10): 1091-112.

[http://dx.doi.org/10.1002/jms.703] [PMID: 15481084]

[94]   La Scola B, Raoult D. Direct identification of bacteria in positive blood culture bottles by matrix-assisted laser desorption ionisation time-of-flight mass spectrometry. PLoS One 2009; 4(11): e8041.
[http://dx.doi.org/10.1371/journal.pone.0008041] [PMID: 19946369]

[95]   Machen A, Drake T, Wang YF. Same day identification and full panel antimicrobial susceptibility testing of bacteria from positive blood culture bottles made possible by a combined lysis-filtration method with MALDI-TOF VITEK mass spectrometry and the VITEK2 system. PLoS One 2014; 9(2): e87870.
[http://dx.doi.org/10.1371/journal.pone.0087870] [PMID: 24551067]

[96]   Barnini S, Ghelardi E, Brucculeri V, Morici P, Lupetti A. Rapid and reliable identification of Gram-negative bacteria and Gram-positive cocci by deposition of bacteria harvested from blood cultures onto the MALDI-TOF plate. BMC Microbiol 2015; 15(1): 124.
[http://dx.doi.org/10.1186/s12866-015-0459-8] [PMID: 26084329]

[97]   Huang AM, Newton D, Kunapuli A, *et al.* Impact of rapid organism identification *via* matrix-assisted laser desorption/ionization time-of-flight combined with antimicrobial stewardship team intervention in adult patients with bacteremia and candidemia. Clin Infect Dis 2013; 57(9): 1237-45.
[http://dx.doi.org/10.1093/cid/cit498] [PMID: 23899684]

[98]   Pliakos EE, Andreatos N, Shehadeh F, Ziakas PD, Mylonakis E. The cost-effectiveness of rapid diagnostic testing for the diagnosis of bloodstream infections with or without antimicrobial stewardship. Clin Microbiol Rev 2018; 31(3): e00095-17.
[http://dx.doi.org/10.1128/CMR.00095-17] [PMID: 29848775]

[99]   Perez KK, Olsen RJ, Musick WL, *et al.* Integrating rapid diagnostics and antimicrobial stewardship improves outcomes in patients with antibiotic-resistant Gram-negative bacteremia. J Infect 2014; 69(3): 216-25.
[http://dx.doi.org/10.1016/j.jinf.2014.05.005] [PMID: 24841135]

[100]   Lockwood AM, Perez KK, Musick WL, *et al.* Integrating rapid diagnostics and antimicrobial stewardship in two community hospitals improved process measures and antibiotic adjustment time. Infect Control Hosp Epidemiol 2016; 37(4): 425-32.
[http://dx.doi.org/10.1017/ice.2015.313] [PMID: 26738993]

[101]   Sharma D, Garg A, Kumar M, Khan AU. Proteome profiling of carbapenem-resistant K. pneumoniae clinical isolate (NDM-4): Exploring the mechanism of resistance and potential drug targets. J Proteomics 2019; 200: 102-10.
[http://dx.doi.org/10.1016/j.jprot.2019.04.003] [PMID: 30953729]

[102]   Sharma D, Garg A, Kumar M, Rashid F, Khan AU. Down-Regulation of Flagellar, Fimbriae, and Pili Proteins in Carbapenem-Resistant *Klebsiella pneumoniae* (NDM-4) Clinical Isolates: A Novel Linkage to Drug Resistance. Front Microbiol 2019; 10: 2865.
[http://dx.doi.org/10.3389/fmicb.2019.02865] [PMID: 31921045]

[103]   Jung JS, Popp C, Sparbier K, Lange C, Kostrzewa M, Schubert S. Evaluation of matrix-assisted laser desorption ionization-time of flight mass spectrometry for rapid detection of β-lactam resistance in Enterobacteriaceae derived from blood cultures. J Clin Microbiol 2014; 52(3): 924-30.
[http://dx.doi.org/10.1128/JCM.02691-13] [PMID: 24403301]

CHAPTER 4

Functional Annotation and Enrichment of Microbial Proteins Using Systems Biology: Tools and Applications

Aditya Arya[*] and **Vivek Dhar Dwivedi**

Centre for Bioinformatics, Computational and Systems Biology, Pathfinder Research and Training Foundation, Greater Noida, Uttar Pradesh, India

Abstract: Genomics and proteomics methods have witnessed a huge surge in the recent decade. Provided the pace of data accumulation from several advanced technologies such as next-generation sequencing, transcriptomics, ChipSeq and quantitative proteomics, the parallel growth in the functional annotation is relatively slow and needs continuous curation and analysis of existed data from repositories. Nevertheless, the standard procedures of functional annotations of proteins which were based on classical data sets, need to be revised in light of advanced and more comprehensive data. In fact, most omics technologies are now integrating and their merger can provide a much realistic and holistic picture of any biochemical and molecular scenario provided the data handling and curation is performed on cellular and molecular principles. A number of genomes and proteomes have been functionally annotated in the past; microbial genomes offer a better opportunity with less complex models. Nevertheless, microbes are among one of the most common causatives of pathogenic diseases and their diagnosis, treatment is limited by available information of proteins and their functions. Previously annotated hypothetical functions to proteins are likely to change in some cases therefore a number of research groups have attempted to re-annotate the microbial genomes with newer data-sets and new tools. Re-annotation of *Mycobacterium tuberculosis* was one such example. Besides pathogenic microbes, emerging trends in various useful microbes sequencing have shown a tremendous increase in information on human microflora and remain a highly prospective area in biology. Systems biology lies at the interphase of biology, mathematics and computational biology and involves a holistic approach to visualize a biological phenomenon. This chapter describes the basic principles involved in the functional annotation of hypothetical proteins in light of emerging datasets and tools. Besides the suggestions on improving standard pipelines, it also presents a summary of recently annotated microbial genomes and future prospects of involving systems biology in the functional annotation for improved quality and output.

[*] **Corresponding author Aditya Arya:** Centre for bioinformatics, computational and systems biology, Pathfinder Research and Training Foundation, Greater Noida, Uttar Pradesh, India; Tel: +91 9811886440; E-mail: contact.adityarya@gmail.com

Divakar Sharma (Ed.)
All rights reserved-© 2020 Bentham Science Publishers

Keywords: Annotations, Microbial proteins, Network enrichment, Systems biology.

INTRODUCTION

The study of microorganisms is important for various applications, such as understanding the diseases caused by them, exploration of their biotechnological potentials and understanding the fundamental processes of life. Nowadays, microorganisms are also being employed for resolving environmental issues and therefore, several non-cultivable forms of microbes are also studied. Among them, various microbes have industrial applications like the production of pharmacologically and commercially important metabolites, the use of microbes as biocatalysts in the food and other industries, and biotransformation of metabolites or pharmacologically important molecules. These applications require a continuous search of microbes with better traits or their improvements by available techniques. Among the classical approaches of improving the microbes is a selection from natural environments or random mutagenesis, while other methods include strain improvement using recombinant DNA technology, mutagenesis or genome editing. Almost all these methods have a prerequisite of genomics, proteomics or metabolomics data and its proper annotation. Advancements in these technologies have greatly advanced the field of microbiology and their industrial applications.

Genome is the complete representative set of genetic material present in an organism and most likely stores the genetic information required for various processes. A large number of genomes of microorganisms have been successfully sequenced and their genomics data is publicly available in repositories (www.nih/genomes). Although genomics information is excellent for the prediction of potential proteins, yet a significant part of the genome is also non-coding, which becomes predominant in higher eukaryotes [1]. Phylogenetics, prediction of putative proteins and identification are important objectives which can be achieved by the study of genomes. Unlike the genome, the proteome, a complete set of proteins in a cell or its compartment, is highly dynamic and remains an important tool for microbial characterization. Metabolome is still more dynamic than proteome and becomes an essential tool for the biotechnological assessment of microorganisms. Several technology upgradation and advancements have led to a prolific rise in the availability of genomics, proteomics and metabolomics data for microorganisms. However, lack of proper annotations and description of proteins in existing databases remains a potential lacuna in the characterization and evaluation of microorganisms and therefore, a number of tools have been developed and annotation remains a continuous process [2].

IMPORTANCE OF ANNOTATION AND ENRICHMENT

Annotation in simple terms is the assignment of a term or value to a variable or a parameter, while enrichment is the global annotation of several variables with a common type of parameter. A more organized definition is "DNA annotation or genome annotation is the process of identifying the locations of genes and all of the coding regions in a genome and determining what those genes do. An annotation (irrespective of the context) is a note added by way of explanation or commentary. Once a genome is sequenced, it needs to be annotated to make sense of it" [3]. For example, if a protein is assigned, its association with a disease will be called annotation and if the disease is assigned to a complete set or a sub-set of proteome within a biological network, it is called enrichment. Enrichment and annotation are highly useful in comparison of genomic and proteomic data and also add value to existing data. Besides this, newly identified genomes may contain some predicted genomic regions and some hypothetical proteins, which may be attributed to functions through the annotation process. An annotated dataset is particularly useful for downstream applications such as drug-discovery, determination of metabolic directions, flux studies and regulation of expression in specific pathophysiological conditions. In the case of microbes, the annotations provide valuable information while developing models for their commercial use and development of useful products. Moreover, previous annotations might help and speed up the genetically related annotation process. Over the past couple of decades, a number of bioinformatics methods have evolved and accumulated in public databases, such as gene ontology, which makes it possible to systematically dissect large gene lists in an attempt to assemble a summary of the most enriched and pertinent biology. Huang *et al.*, in their review, described a number of high-throughput enrichment tools, including, but not limited to, Onto-Express, MAPPFinder, GoMiner, DAVID, EASE, GeneMerge and FuncAssociate, *etc.*, independently developed during 2002 and 2003 as initial studies to address the challenge of functionally analyzing large gene lists. Since then, the enrichment analysis field has been very productive, resulting in more similar tools becoming publicly available [4].

EMERGING TRENDS IN GENOMICS

Since the advent of DNA sequencing in the early 1960s, the genomes of microorganisms begun to be sequenced and that led to the quest of human genome sequencing, which was published as an initial draft in the year 2001 [5 - 7] and later a complete draft in 2003. With this, the rapid proliferation of genome sequencing begun and also the technology improved to a great extent. Conventional Sanger's sequencing modified a lot over a few decades and the emergence of next-generation sequencing revolutionized the whole paradigm.

Over thousands of microbial genomes have been completed and a newer insight in various genomics components has descended (Fig. **1**) illustrates the cumulative number of microbial genome sequenced so far.

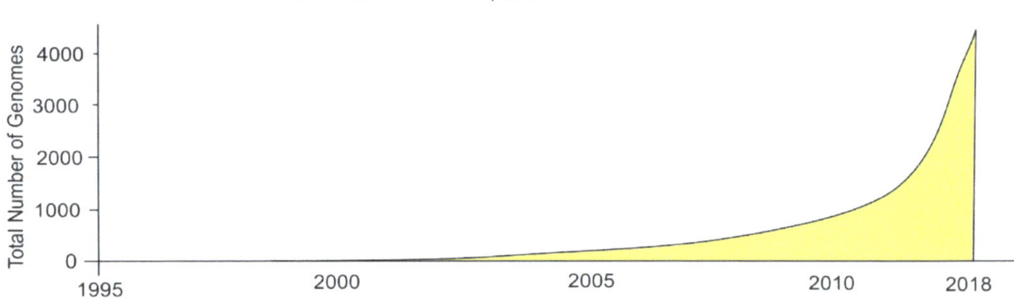

Fig. (1). Overview of the number of microbial genomes sequenced over a period of years.

At present, several newer approaches of sequencing named next-generation sequencing have emerged, which has paced up the process of genome sequencing. The initial advent of next-generation sequencing occurred between 2006 and 2009 with the emergence of pyrosequencing technology, but could not exist longer as much better platforms dominated. Technologies such as sequencing by ligation and sequencing by synthesis are a popular choice, with illumine platforms being widely used. The cost and time consumed for a whole-genome sequence have substantially reduced to approximately 0.1 USD per base and 1 day per genome, respectively and still growing. Third generation methods such as nanopore technology developed by oxford have become a very handy and portable next-generation sequencing platform, especially for fieldwork, genome sequencing at remote sites or sequencing of uncultivable microbes.

Recently, whole-genome assembly of European *Aspergillus terreus* clinical isolate M6925 has been derived by single-molecule real-time sequencing with short read polishing [8]. The draft genome of LSUCC0115, a novel coastal Gulf of Mexico bacterioplankton isolate, has been sequenced using IlluminaMiSeq technology [9]. A complete genome sequence was obtained through the illuminaMiSeq system for a severe acute respiratory syndrome coronavirus 2 (SARS-CoV-2) strain isolated from an oropharyngeal swab specimen of a Nepalese patient with coronavirus disease 2019 (COVID-19), who had returned to Nepal after traveling to Wuhan, China [10].

EMERGING TRENDS IN PROTEOMICS

Proteomics is much dynamic and also dictates the physiological state of an organism or cell, or subcellular cell compartment. In the early days, proteomic methods such as immunoblotting and other immune reactive strategies became popular for the identification and diagnosis of various microbes. Moreover, gel electrophoresis is an important tool in proteomics, which allowed an efficient separation of proteins in a complex mixture, yet the identification was a tricky process which was resolved by the invention of mass spectrometry, which later became a popular choice of proteomic biologists. Mass spectrometry in the early days was followed by two-dimensional gel electrophoresis and the usual version used for mass spectrometry was MALDI-TOF, *i.e.*, matrix assisted laser desorption ionization-time of flight, where proteins were trypsinized, mixed with an organic matrix and allowed to be ionized by laser followed by separation of fragments on the basis of charge by mass ratio. In 1989, half of the Nobel Prize in Physics was awarded to Hans Dehmelt and Wolfgang Paul for the development of the ion trap technique in the 1950s and 1960s. In 2002, the Nobel Prize in Chemistry was awarded to John Bennett Fenn for the development of electrospray ionization (ESI) and Koichi Tanaka for the development of soft laser desorption (SLD) and their application to the ionization of biological macromolecules, especially proteins. (Fig. **2**) depicts some of the classical approaches used in proteomics.

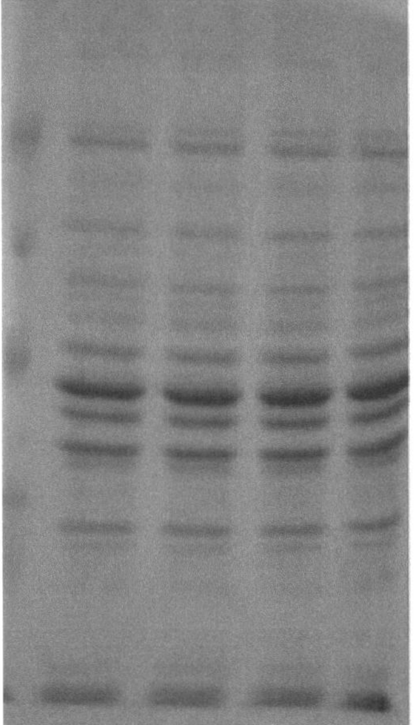

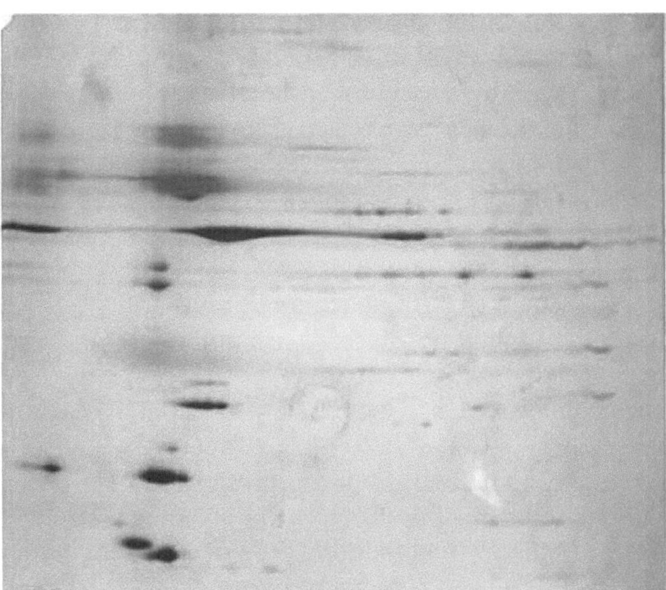

Fig. (2). A typical one-dimensional and two-dimensional gel illustrating the separation of proteins.

Two-dimensional gel electrophoresis methods followed by MALDI-TOF have long been challenged for reproducibility and therefore reliability and use gradually taken over by newer methods in proteomics [11, 12]. At present, most proteomics are performed using advanced proteomics platforms which integrate the protein separation, such as liquid chromatography–mass spectrometry [13, 14]. Mass spectrometry has also emerged with quantitative measurements and multiplexing of several samples by using various approaches such as iTRAQ (isobaric tags), ICAT, label free system, Mudpit and Swathe. Moreover, the nature of data generated from quantitative systems is much wider and extensive compared to previous methods and thus provides us with an opportunity to achieve better annotation at the next level of data processing.

A strategy based on high-throughput peptide synthesis and mass spectrometry was used to generate an almost complete reference map (97% of the genome-predicted proteins) of the *Saccharomyces cerevisiae* proteome [15]. A recent orthogonal proteomic study identifies Zikavirus (ZIKV) host-dependency factors and provides a comprehensive framework for a system-level understanding of ZIKV-induced perturbations at the levels of proteins and cellular pathways [16].

ANNOTATION OF MICROBIAL PROTEINS

Most of the time, high throughput techniques which generate a huge amount of data, such as next-generation sequencing, microarray, mass-spectrometry do not provide enough information about the assignment of a function to a gene or protein. Annotations often follow these methods, which give them their identity and relation with the biochemical function. There are several classical approaches for the annotation of microbial proteins. Most common annotations include predictions of hypothetical predictions in newly sequenced or partially understood genomes. Protein, whose existence has been predicted, but lacks the experimental evidence of its presence *in vivo*, is commonly known as a hypothetical protein. Sequencing of several genomes has resulted in numerous predicted open reading frames to which functions cannot be readily assigned. Scientific literature suggests that orphan or conserved hypothetical proteins make up nearly 20% to 40% of proteins encoded in each newly sequenced genome. Nowadays, most protein sequences are inferred from computational analysis of genomic DNA sequences. Hypothetical proteins are created by gene prediction software during genome analysis. When the bioinformatics tool used for the gene identification finds a large open reading frame without a characterized homologue in the protein database, it returns "hypothetical protein" as an annotation remark.

We and others previously worked on re-annotation of the *Mycobacterium tuberculosis* genome [17]. In this study, the original genome of *Mycobacterium tuberculosis* was re-evaluated and unannotated proteins were filtered out. The protein sequences of such proteins either derived from the genome or directly from UniProt were evaluated using several servers such as ProDom – used for identification of proteins, Pfam- for the identification of protein families. Furthermore, structural and sequence alignment was performed to assess the nearest neighbors and proteins were then annotated. This was interestingly performed *via* community based effort and proven to be a successful strategy [18], referred to as an open-source drug discovery program (OSDD).

SYSTEMS BIOLOGY AND ANNOTATIONS

Systems biology is a recent domain added to biology with close interaction with an overlap with several other domains such as mathematics, physics, chemistry and computer sciences. Although there is not a common consensus on a unified definition of systems biology considering its highly interdisciplinary nature and varied applications. Ron Germain does have his own definition of systems biology that he's sticking to: a scientific approach that combines the principles of engineering, mathematics, physics, and computer science with extensive experimental data to develop a quantitative as well as a deep conceptual understanding of biological phenomena, permitting prediction and accurate simulation of complex (emergent) biological behaviors (NIAID, NIH). Specialists from various domains involved in systems biology make use of different tools for the exploration of proteome, genome or metabolome. In the context of microbial genomics and proteomics, mostly quantitative is often used in the systems biology for the analysis using systems biology tools. Among the common approaches is the use of network biology, which has recently gained attention.

Biological networks are of various types, such as protein interaction networks, canonical pathway networks, molecular networks, *etc.* Each of these networks can be further enriched with various diseases, gene ontology and a large number of features.

A large number of tools are routinely used for the enrichment of various omics data for annotations and further analysis. Among one of the most commonly used protein interaction databases and tools is STRING [19, 20], which can provide not only information about the interacting partners of a protein but also their associated partners (Fig. **3**). Analyzing a larger clad and clustering algorithms provides additional information about the regulatory pathways of microbes, such as the establishment of upstream regulators and hub or master regulators of a common biological process. Besides STRING, another commonly used open-source platform is Cytoscape [21]. This tool is based on java and enables its users to analyze a wide variety of features and visualization of the data in a much flexible manner. Besides Cytoscape, various other tools, both proprietary and open-source, are used for microbial annotations. Table **1** Enlists some of the tools used in systems biology for microbial functional annotations.

Table 1. Commonly used tools for systems biology.

Name of Database	Weblink (Download or Access)	Accessibility	Year of Release	Key Features	Type of Tool	Developers Name (Optional)	User Friendly
Gephi	https://gephi.org/	Open Source	2008	Data import from databases, Network topology analysis, high quality image export, Annotation and editing	Both	Mathieu Bastian, Eduardo Ramos Ibañez, Mathieu Jacomy, CezaryBartosiak, SébastienHeymann, Julian Bilcke, Patrick McSweeney, André Panisson, JérémySubtil, Helder Suzuki, Martin Skurla, Antonio Patriarca	4
Proviz	http://proviz.ucd.ie/	Open Source	2004	Data import from databases	Standalone	Iragne	4
NodeXL	http://nodexlgraphgallery.org/Pages/Default.aspx	Open Source	2008	Data import from databases, Multiple layouts, Predictions, Network topology analysis, high quality image export, Annotation and editing	Standalone	Social Media Research Foundation	5
Pajek	http://pajek.imfm.si/doku.php	Open Source		Data import from databases, high quality image export	Online	Vladimir Batagelj, Andrej Mrvar	4
ACGT	http://acgt.cs.tau.ac.il/pivot/)	Open Source		Data import from databases, Predictions, Network topology analysis, high quality image export, Annotation and editing	Online	NirOrlev Prof. Yossi Shiloh and Prof. Ron Shamir with the assistance of Giora Sternberg	4
Patika	http://www.cs.bilkent.edu.tr/~patikaweb/	Open Source		Data import from databases, high quality image export, Annotation and editing	Online		4

(Table 1) cont.....

Name of Database	Weblink (Download or Access)	Accessibility	Year of Release	Key Features	Type of Tool	Developers Name (Optional)	User Friendly
Ondex	https://sourceforge.net/p/ondex/wiki/Home/	Open Source		Data import from databases, high quality image export, Annotation and editing	Standalone		4
Sprey	https://osprey.thebiogrid.org/	Open Source		Data import from databases, Multiple layouts, Network topology analysis, high quality image export, Annotation and editing	Standalone	TyersLab	3
Medusa3	https://sites.google.com/site/medusa3visualization/	Open Source	2002	Data import from databases, Multiple layouts, Annotation and editing	Standalone	CAD Schroer	3
Graphia	https://kajeka.com/graphia-professional/	Open Source		Multiple layouts, Predictions, Network topology analysis, high quality image export, Annotation and editing	Both	Prof Tom Freeman (University of Edinburgh) and Dr Anton Enright (Sanger Institute/EBI)	4
Jung	http://jung.sourceforge.net/team.html	Open Source	2003	Multiple layouts, Predictions, high quality image export, Annotation and editing	Both		3
PINA	https://omics.bjcancer.org/pina/home.do	Open Source		Data import from databases, Network topology analysis, high quality image export	Online		3

(Table 1) cont.....

Name of Database	Weblink (Download or Access)	Accessibility	Year of Release	Key Features	Type of Tool	Developers Name (Optional)	User Friendly
IPA	https://analysis.ingeneuity.com/pa/installer/selecy	Propitiatory	2014	Data import from databases, Multiple layouts, Predictions, Network topology analysis, high quality image export, Annotation and editing	Standalone	QIAGEN	5
Pajek	http://mrvar.fdv.uni-lj.si/pajek/	Open Source	2011	Data import from databases, Predictions, Network topology analysis, high quality image export, Annotation and editing	Online		4
String	https://string-db.org/	Open Source	2002	Data import from databases, Predictions, high quality image export, Annotation and editing	Online		5
Cytoscape	https://cytoscape.org/	Open Source	2002	Data import from databases, Multiple layouts, Network topology analysis, high quality image export, Annotation and editing	Standalone	Institute for system biology	4
UCINET	https://sites.google.com/site/ucinetsoftware/home	Propitiatory	2002	Data import from databases, Network topology analysis, high quality image export	Standalone	Lin freeman,MartinEverett,SteveBorgatti	3

(Table 1) cont.....

Name of Database	Weblink (Download or Access)	Accessibility	Year of Release	Key Features	Type of Tool	Developers Name (Optional)	User Friendly
Netminer	http://www.netminer.com/main/main-read.do	Propitiatory	2001	Data import from databases, Multiple layouts, Predictions, Network topology analysis, high quality image export, Annotation and editing	Standalone	Cyram Inc.	4
SocenTV	https://socnetv.org/	Open Source	2016	Data import from databases, Multiple layouts, Predictions, Network topology analysis, high quality image export, Annotation and editing	Online		4
Solarwind	https://www.solarwinds.com/network-performance-monitor/use-cases/network-visualization	Propitiatory		Data import from databases, Multiple layouts, Predictions, Network topology analysis, high quality image export, Annotation and editing	Standalone		3
Graphviz	https://www.graphviz.org/	Open Source	1991	Data import from databases, Multiple layouts, Predictions, Network topology analysis, high quality image export, Annotation and editing	Standalone	AT&T Labs Research	4
Commetrix	http://www.commetrix.de/	Open Source	1997	Network topology analysis	Standalone		4
Cuttlefish	http://cuttlefish.sourceforge.net/	Open Source	2007	Network topology analysis	Online	Chair of Systems Design of ETH Zürich	4

(Table 1) cont.....

Name of Database	Weblink (Download or Access)	Accessibility	Year of Release	Key Features	Type of Tool	Developers Name (Optional)	User Friendly
Metacore	https://portal.genego.com/	Propitiatory	2008	Network topology analysis	Standalone	Thomson Reuters	5

In a recent study, novel coronavirus 2019-nCoV/SARS-CoV-2 drugs have been repurposed using a systems biology network-based approach [22]. Kundu *et al.* in 2009 analyzed the species-wide metabolic interaction network for understanding the natural lignocellulose digestion in termite gut microbiota [23]. The close relationship between bacterial metabolism and antibiotic efficacy was emphasized using network analysis [24].

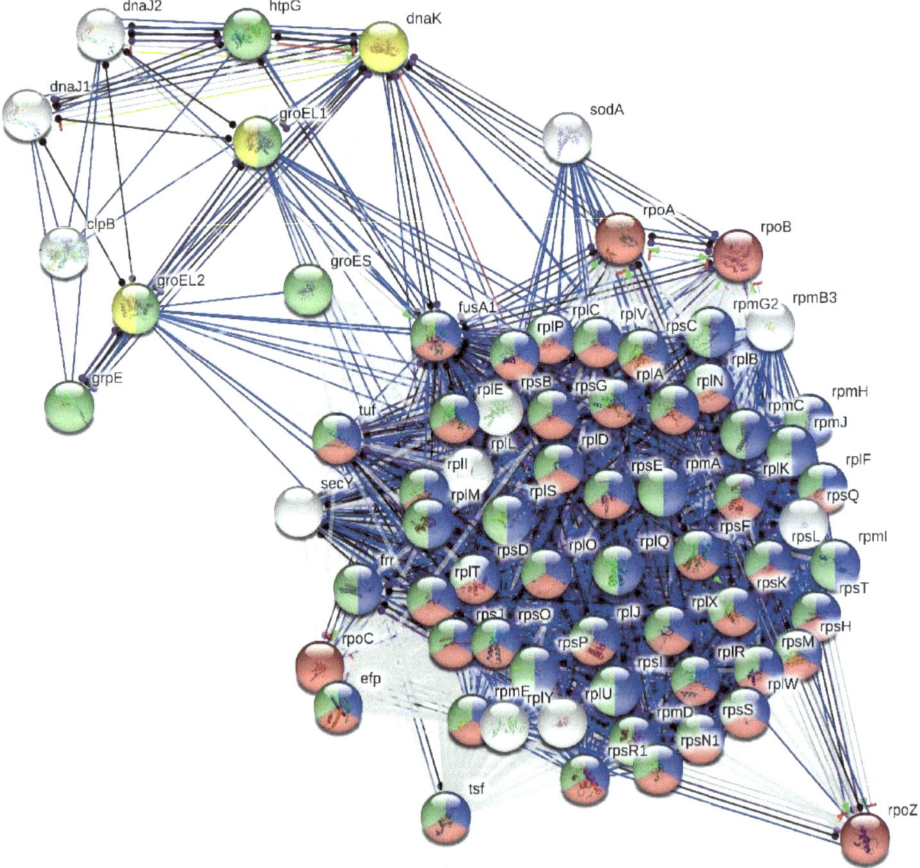

Fig. (3). A typical protein interaction network and its enrichment using STRING. Molecular action based interaction of GroES protein of *Mycobacterium tuberculosis*; Enriched with gene ontology terms (metabolic process – blue, nucleic acid binding – red, cytoplasmic component –green, KEGG pathway of RNA degradation).

FUTURE GUIDELINES

With a rapid ascension of omics technologies and the amount of biological data available, the existing and conventional methods of functional annotation and enrichment are becoming time-consuming and less reliable. Availability of effective visualization and computer algorithms for analysis of systems biology provides much better predictions, which are increasingly being used. In the future, it is anticipated that multi-omics platforms integrating the cross-platform data, cross-organism data shall provide a much better and comprehensive understanding of microbial systems and predictions with better statistical confidence. In fact, machine learning and deep learning strategies are soon expected to plunge into this domain and likely to improvise them to a great extent.

CONSENT FOR PUBLICATION

Not applicable.

CONFLICT OF INTEREST

The author declares no conflict of interest, financial or otherwise.

ACKNOWLEDGEMENTS

Declared none.

REFERENCES

[1]   Fraser CM, Eisen JA, Salzberg SL. Microbial genome sequencing. Nature 2000; 406(6797): 799-803.
        [http://dx.doi.org/10.1038/35021244] [PMID: 10963611]

[2]   Gabanyi MJ, Berman HM. Protein structure annotation resources. Structural Proteomics. New York, NY: Humana Press 2015; pp. 3-20.

[3]   Cseke LJ, Kirakosyan A, Kaufman PB, Westfall MV, Eds. Handbook of molecular and cellular methods in biology and medicine. CRC press 2011.

[4]   Huang W, Sherman BT, Lempicki RA. Bioinformatics enrichment tools: paths toward the comprehensive functional analysis of large gene lists. Nucleic Acids Res 2009; 37(1): 1-13.
        [http://dx.doi.org/10.1093/nar/gkn923] [PMID: 19033363]

[5]   Sanger F, Nicklen S, Coulson AR. DNA sequencing with chain-terminating inhibitors. Proc Natl Acad Sci USA 1977; 74(12): 5463-7.
        [http://dx.doi.org/10.1073/pnas.74.12.5463] [PMID: 271968]

[6]   Venter JC. A part of the human genome sequence. Science 2003; 299(5610): 1183-4.
        [http://dx.doi.org/10.1126/science.299.5610.1183] [PMID: 12595674]

[7]   Lander ES, Linton LM, Birren B, *et al.* International Human Genome Sequencing Consortium. Initial sequencing and analysis of the human genome. Nature 2001; 409(6822): 860-921.
        [http://dx.doi.org/10.1038/35057062] [PMID: 11237011]

[8]   Palanivel M, Mac Aogáin M, Purbojati RW, *et al.* Whole-Genome Sequencing of *Aspergillus terreus* Species Complex. Mycopathologia 2020; 185(2): 405-8.

[PMID: 32108289]

[9]   Henson MW, Guidry ME, Carnes MK, Thrash JC. Draft genome sequence of the novel coastal bacterium LSUCC0115 from the MWH-UniPo Clade, Order Burkholderiales, Class Betaproteobacteria. Microbiology Resource Announcements 2020 Jan 9; 9(2): e01492-19.

[10]   Sah R, Rodriguez-Morales AJ, Jha R, *et al.* Complete genome sequence of a 2019 novel coronavirus (SARS-CoV-2) strain isolated in Nepal. Microbiology Resource Announcements 2020 Mar 12; 9(11): e00169-20.

[11]   Sharma D, Lata M, Singh R, Deo N, Venkatesan K, Bisht D. Cytosolic proteome profiling of aminoglycosides resistant *Mycobacterium tuberculosis* clinical isolates using MALDI-TOF/MS. Front Microbiol 2016; 7: 1816.
[http://dx.doi.org/10.3389/fmicb.2016.01816] [PMID: 27895634]

[12]   Sharma D, Bisht D. Secretory proteome analysis of streptomycin-resistant *Mycobacterium tuberculosis* clinical isolates. 2017 Dec; 22(10): 1229-38.
[http://dx.doi.org/10.1177/2472555217698428]

[13]   Sharma D, Garg A, Kumar M, Khan AU. Proteome profiling of carbapenem-resistant K. pneumoniae clinical isolate (NDM-4): exploring the mechanism of resistance and potential drug targets. J Proteomics 2019 Dec May 30; 200: 102-10.

[14]   Sharma D, Garg A, Kumar M, Rashid F, Khan AU. Down-regulation of flagellar, fimbriae, and pili proteins in carbapenem-resistant klebsiellapneumoniae (ndm-4) clinical isolates: a novel linkage to drug resistance. Frontiers in Microbiology 2019; 10.

[15]   Picotti P, Clément-Ziza M, Lam H, *et al.* A complete mass-spectrometric map of the yeast proteome applied to quantitative trait analysis. Nature 2013; 494(7436): 266-70.
[http://dx.doi.org/10.1038/nature11835] [PMID: 23334424]

[16]   Scaturro P, Stukalov A, Haas DA, *et al.* An orthogonal proteomic survey uncovers novel Zika virus host factors. Nature 2018; 561(7722): 253-7.
[http://dx.doi.org/10.1038/s41586-018-0484-5] [PMID: 30177828]

[17]   Anand P, Sankaran S, Mukherjee S, *et al.* OSDD Consortium. Structural annotation of *Mycobacterium tuberculosis* proteome. PLoS One 2011; 6(10): e27044.
[http://dx.doi.org/10.1371/journal.pone.0027044] [PMID: 22073123]

[18]   Bhardwaj A, Scaria V, Raghava GP, *et al.* Open Source Drug Discovery Consortium. Open source drug discovery--a new paradigm of collaborative research in tuberculosis drug development. Tuberculosis (Edinb) 2011; 91(5): 479-86.
[http://dx.doi.org/10.1016/j.tube.2011.06.004] [PMID: 21782516]

[19]   Szklarczyk D, Gable AL, Lyon D, *et al.* STRING v11: protein-protein association networks with increased coverage, supporting functional discovery in genome-wide experimental datasets. Nucleic Acids Res 2019; 47(D1): D607-13.
[http://dx.doi.org/10.1093/nar/gky1131] [PMID: 30476243]

[20]   Sharma D, Singh R, Deo N, Bisht D. Interactome analysis of Rv0148 to predict potential targets and their pathways linked to aminoglycosides drug resistance: An insilico approach. Microb Pathog 2018; 121: 179-83.
[http://dx.doi.org/10.1016/j.micpath.2018.05.034] [PMID: 29800702]

[21]   Shannon P, Markiel A, Ozier O, *et al.* Cytoscape: a software environment for integrated models of biomolecular interaction networks. Genome Res 2003; 13(11): 2498-504.
[http://dx.doi.org/10.1101/gr.1239303] [PMID: 14597658]

[22]   Zhou Y, Hou Y, Shen J, Huang Y, Martin W, Cheng F. Network-based drug repurposing for novel coronavirus 2019-nCoV/SARS-CoV-2. Cell Discov 2020; 6(1): 14.
[http://dx.doi.org/10.1038/s41421-020-0153-3] [PMID: 32194980]

[23]   Kundu P, Manna B, Majumder S, Ghosh A. Species-wide Metabolic Interaction Network for

Understanding Natural Lignocellulose Digestion in Termite Gut Microbiota. Sci Rep 2019; 9(1): 16329.
[http://dx.doi.org/10.1038/s41598-019-52843-w] [PMID: 31705042]

[24]   Stokes JM, Lopatkin AJ, Lobritz MA, Collins JJ. Bacterial metabolism and antibiotic efficacy. Cell Metab 2019; 30(2): 251-9.
[http://dx.doi.org/10.1016/j.cmet.2019.06.009] [PMID: 31279676]

CHAPTER 5

Emerging Paradigm of Post-translational Modifications in Microbes: An Undefeatable Weapon

Arpana Sharma[1,2], Chandrajeet Singh[1], Gopika Raval[2], Kruti Dave[2], Ankita Mathur[3], Juhi Sharma[4] and Divakar Sharma[5,*]

[1] *Central University of Gujarat, Gandhinagar, India*

[2] *Mehsana Urban Institute of Sciences, Department of Biotechnology, Ganpat University, Kherva, 384012, India*

[3] *Mewar University, Gangrar, Chittorgarh, Udaipur, 312901, India*

[4] *Department of Botany and Microbiology, St. Aloysius College, Sadar Cantt, Jabalpur, 482002, India*

[5] *CRF, Mass Spectrometry Laboratory, Kusuma School of Biological Sciences (KSBS), Indian Institute of Technology, Delhi, 110016, India*

Abstract: Bacteria are simple organisms, optimized with basic but robust cell regulation mechanisms for efficient growth. Nonetheless, fitting in the description, they possess remarkable adaptive capacity with diverse survival strategies. Post-translational modifications (PTMs) provide a competitive edge to adapt bacteria with limiting, fluctuating nutrients and extremes of environment variables. PTM is one of the cellular processes in bacteria and helps them to adapt to a new environment for their survival. Several post-translation modifications regulate bacterial functions and provide strength to bacteria for surviving in adverse conditions. On-going investigations revealed many reversible or irreversible novel bacterial PTMs like the addition of simple group (acetylation, phosphorylation, methylation and hydroxylation) or composite molecules (AMPylation, ADP-ribosylation, glycosylation and isoprenylation), or small protein ubiquitin and modifying the side chain residues like (elimination and deamidation). PTMs are recognized as important players in directing cellular dynamics like cell metabolism, stress response, pathogenesis and virulence factors. Bacteria with several PTMs are capable of modulating the signaling pathways by destabilising the host cell defense machinery, protein-protein interactions, ultimately promoting their replication. Currently, many studies focus on the relationship between PTMs and antibiotic resistance, increasing bacterial tolerance to various antimicrobials. A paradoxical behaviour is that a single protein may be modified at one or variable positions in interspecies, but changes also exist in the same species. So, to characterize the multifaceted interactions of PTMs, it is still a challenge in metabolic engineering,

* **Corresponding author Divakar Sharma:** CRF, Mass Spectrometry Laboratory, Kusuma School of Biological Sciences (KSBS), Indian Institute of Technology, Delhi, 110016, India; Tel: +91-7906026680; E-mail: divakarsharma88@gmail.com

Divakar Sharma (Ed.)
All rights reserved-© 2020 Bentham Science Publishers

synthetic biology, and medical sciences. Therefore, understanding of bacterial PTMs and PTMs directed host proteins modification will provide better insights into host-pathogen interaction. This chapter focuses on the roles of PTMs in nutrition sequestration and cellular response. Furthermore, we discuss the prospects and advances of proteomics tools in enhancing knowledge related to PTMs of human gut microbiota.

Keywords: Acetylation, Carboxylation, Glycosylation, Interactomics, Lipidation, Methylation, Microbial proteins, Nitrosylation, Phosphorylation, Post-translational modifications, Proteomic tools.

INTRODUCTION

Bacteria are simple, single-celled organisms that can thrive in extreme environmental conditions. They are classified as prokaryotic cells which have a simple internal structure, lack a nucleus and contain DNA in the form of a loosely warped fiber-like structure called "nucleoid". The relationship of human-bacteria is complicated as some bacteria are useful while others are disease-causing. They have been seen in many associations such as symbiotic, parasitic or commensalism. In the symbiotic associations, natural selection favours the organisms to interact within species in a manner to favour them in extreme environments, which is absent in the individual species. The bacteria, which associate together as co-cultures, which have many advantages in the biotechnology industries, are also a very important part of gut microbiota, which are essential for the normal working of the host organisms [1]. Most of the mutualistic associations are harmless, but there are some which are harmful and cause the disease to the human species. Such bacteria are identified as pathogenic organisms and their symbiotic interactions cause infectious diseases and host damage [2, 3]. These pathogenic microorganisms adapt themselves by immune-compromising the host by suppressing innate immunity to survive in the host cells, subsequently posing a constant challenge to human wellbeing.

Hence, infection biology has been in use over the years to study the abnormal responses which are acquired by the pathogenic organisms inside the host cells. The bacteria have a specific infection cycle, which is identified [4]. Firstly they will introduce into the host tissue by a specific mode of transmission. Subsequently, they will adhere to the tissues and establish themselves by evading the host immune responses. Consequently, they will disseminate into the nearby tissues by secreting toxins. In the process, they will hijack the host molecular mechanisms and mimic the same for its growth.

Nowadays, there is an appearance of many multidrug resistance species, which has created a lot of curiosity in the infection biology fields. Also, with the advancement of proteomic techniques, there is a lot of research going on to study the molecular signals during infections. The assertiveness of virulence and infection is due to the secretion of various proteins that may be secreted in the original form or may be modified [5].

It has been investigated that various post-translational modifications have been studied, which have been shown to increase the virulence and pathogenicity of bacteria. It is pivotal that these protein secretion systems must be studied and understood for the eradication of the pathogens. To curb the bacterial disease, it is important to target any of these proteins and their secretion systems [6].

Bacterial cells change their endogenous biochemical functions to adapt to the altering environmental conditions. They achieve these adjustments by successive post-translational modifications (PTMs), which alter the biological and chemical properties of proteins. Till date, approximately 200 PTMs have been established and illustrated by many novel proteomic approaches. These PTMs are very important as they confer important protein function properties, maintains its stability, modifies its activity to the changing environmental cues and helps in establishing interactions with other proteins [7 - 11].

The chapter will focus on the different types of post-translational systems involved in bacterial pathogenicity, their implication in providing virulence to pathogens, and how they help in establishing the host-pathogen interactions. The following sections will also elucidate the various proteomics tools and technologies available to study these post-translational modifications and what are the technical difficulties involved in studying such systems.

What are the General Characteristics of Bacteria and the Significance of the Study?

Bacteria are simple but robust cellular entities that have basic cellular machinery, which is better developed for speedy division and rapid growth. But, bacteria, despite being uncomplicated, acquire special properties of adaptations to survive in diverse extreme situations by modifying their size, morphology, shape development patterns and metabolism behaviour [12]. This robust cellular mechanism and amazing adaptive capacity have given emergence to certain pathogenic strains of bacteria. The flora which is present in healthy individuals is essential for accurate working but multiple pathogenic organisms can mutually coexist in the same host niche to produce unnatural responses [12].

These responses have to be studied in the co-existence of multiple microbial species conditions, which break the immune system of the host and elicit virulence conditions. Hence, bacterial co-culture growth analysis in an appropriate manner is very important to study the microbial behaviour causing infections. With the advancement of proteomic tools, mass spectroscopy, improvements in the high-resolution devices and progress in experimental techniques have helped expand the scope of infection biology [4].

A myth that has been destroyed from the past is that PTMs are only confined to simple bacteria and missing in the complex organisms. But recently, it has been confirmed that, PTMs are ubiquitously present and very important contemplation for studying infection biology. Bacteria modify their proteins with simple chemical alterations, which help them to adapt to extreme environments. The PTMs are very dynamic, reversible regulatory molecules, which are exploited by the microorganisms to survive the normal and harsh environments [12]. These PTMs are very significant, as they modulate the cell cycles to give a survival edge to the bacteria. So to study the cellular roles of these bacteria, many proteomic tools have been used to study the quantitative expression of the proteome, one such tool being Intensity Based Absolute Quantitation (iBAQ) [13]. These proteomic tools also helped us in the establishment of novel PTMs which are involved in the pathogenicity of organisms.

A group of disease-causing organisms called "ESKAPE" bacteria originated, which escape the antibiotic treatment strategies. They have been recognized by the Infectious Diseases Society of America (IDSA), which include Enterococcus, Staphylococcus, Klebsiella, Pseudomonas and many more. They have developed novel ways of transmission and evading the immune system in the host, demonstrating the emergence of resistant bacteria [14 - 17]. These antibiotic resistant microorganisms developed due to the exploitative use of antibiotics cause a serious threat to human healthcare, posing continuous danger to life [18]. There are many strategies that have been employed to explore the resistance and possible treatment of these resistant bacteria like nano-medicine and herbal formulations but none of the alternatives are best suitable availability to treat these ESKAPE pathogenic organisms [19 - 22].

Therefore, it is evident to search for new drug targets which may be the best option for treating these diseases efficiently and rapidly decreasing the drug resistance, so it is a prime concern that we understand all the factors which are included in the virulence transmission, survival and adaptive mechanisms of these organisms growth even during the hostile conditions [23]. Some of the important new pharmacological targets may be proteins involved in adhesion, quorum sensing, enterotoxin secretion, which might be new anti-virulence mediators helpful in disarming the bacteria [24 - 26]. In recent times, there is a trend of discovering new antibacterials to target the multi-resistant disease-causing bacteria; hence the role of bacterial PTMs is important to be investigated, which may be helpful to serve as potential targets.

What are the Post-translation Modifications?

"Post-translational modifications" (PTMs) are reversible simple covalent modifications where simple or covalent chemical groups are attached to specific amino acids, after the polypeptide synthesis (translation) and during protein folding [27]. Some specific chemical alterations like amino-terminal acetylation or glycosylation take place during the translational process simultaneously; hence, some modifications are co-translational [28, 29]. In nature, we have only 20 amino-acids which commonly form proteins, but due to these modifications, the cells extend their range of information of protein synthesis to nearly 140 amino-acids, which can constitute proteins [30]. Hence, prokaryotes use these amazing weapons called PTMs, which offer extremely adaptable versatile tools to normalize the action of crucial proteins inside the cells. Fig. (**1**) shows the occurrence of the most common PTMs in bacteria [31].

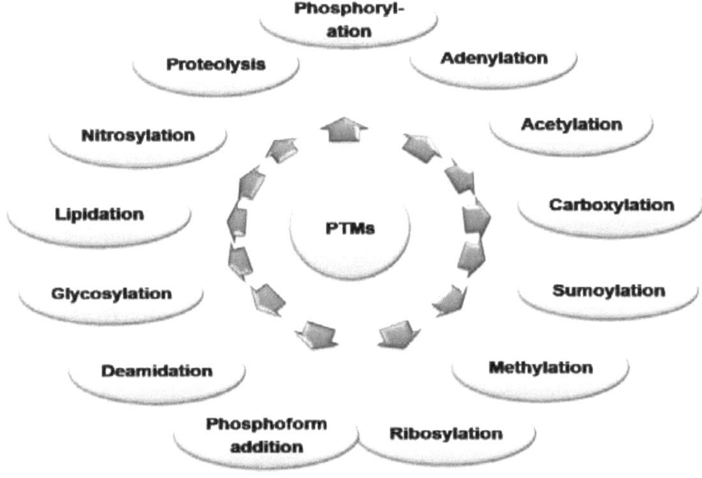

Fig. (1). PTMs present in the bacterial proteome.

PTMs have the ability to influence the chemical characteristics of amino-acids residues, which are modified in addition to the adjacent polypeptide chains, which help in determining the higher degree of protein conformation like tertiary, quaternary structural features, charge, binding with neighbouring proteins, localisation, their stability and degradative cycles and ultimately their functions [27, 32]. The following figure shows the range of chemical group modifications occurring in the bacterial proteins (Fig. **2**) and the most important amino-acids frequently undergoing PTMs (Fig. **3**).

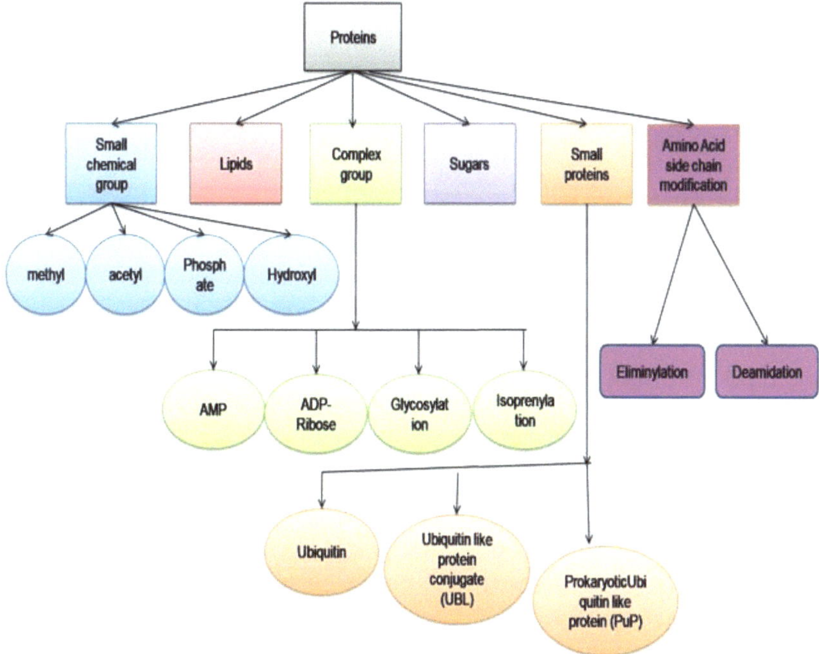

Fig. (2). Different groups of PTMs present in proteins.

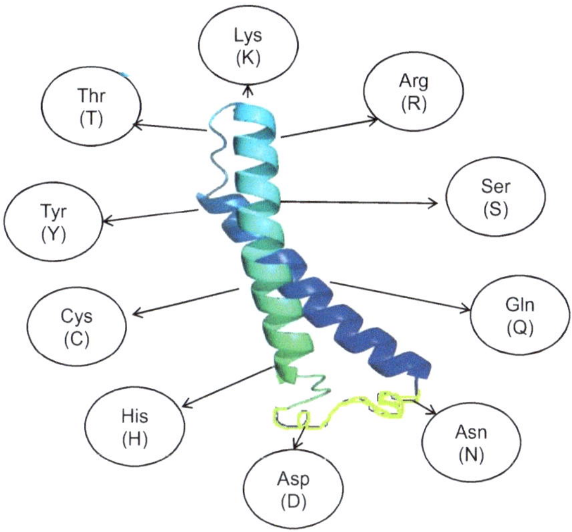

Fig. (3). Amino-acids residues getting modified in PTMs.

The chemical modifications occurring in bacterial proteins are mostly reversible and dynamic, occurring in time with a signal-dependent approach. PTMs are unique substoichiometric features of only known proteins, which is a remarkable feature. They are made on the polypeptide chains with the help of enzymes or may be spontaneously occurring depending on the signaling [33]. Enzymes-related modifications are usually very specific and tightly regulated as they can modify their target substrates. These substrates are responsible for the signal transduction pathways and enzymes are key molecules for the regulation of the PTMs which are involved in these cascade mechanisms [27]. This enables the modifying enzymes of the cells to do the energetically favorable alterations on the specific proteins very easily, rapidly and without consumption of any expensive protein degrading mechanisms [34]. Conversely, irreversible nonenzymatic driven PTMs like thiolation and acetylation are also present. Many thiolation protein modifications are caused due to reactive specics and free radicals like oxygen, chlorine; hence they are important in regulating redox potential [35]. They help in activating specific transcription factors that express the specific genes involved in detoxification pathways [36].

Bacterial modifications are tedious to study because of the enormity of diversity present and not very frequent occurrence makes it a cumbersome task. In order to study the post-translational modifications in bacterial proteins, the requirement of a high amount of starting material together with good proteomic tools, biochemical analysis workflows, highly developed biochemical tools is a must [37]. Rapid progress has been made with the recent advancement in proteomics, high-resolution mass spectrometry, 2-D gel electrophoresis, and improved bioinformatics datasets, which have helped decipher the novel PTMs in many bacterial species [38]. Using these strategies, the first best-elaborated study of protein phosphorylation was conducted on *Escherichia coli* phosphoproteome, which was documented in 2008. They studied 81 phosphorylation sites on histidine, but recently the number has tremendously increased to 1,883 with more sophisticated tools [39]. Another breakthrough was achieved with the *Bacillus subtilis* and *Staphylococcus aureus* gram-positive model organisms, where hundreds of novel arginine phosphorylation sites have been identified [40, 41]. All this technical progress has helped the microbiologists to detect other PTMs like lysine acetylation, glycosylation, *etc.* and analyze their biological significances.

However, PTMs studies are surrounded with many technical challenges, and so there are some disadvantages too. In the *in vivo* conditions, only 1% of the proteins are modified. So it becomes very difficult during analyzing when modified and unmodified mixtures of proteins are used for detection. As current proteomic technology has the capability of recognizing only simple alterations from the heterogeneous mixtures, the next challenge is to study the native conditions of the proteins because when sample preparations are carried, the probability of disruption of proteins increases, which may hinder the study of endogenous modifications of proteins [33].

Different bacterial PTMs are expressed in varying conditions; many reports show a diverse number of PTMs, which have been documented to have different expression and residency sites in different environmental growth parameters. The present proteomics data set has shown around 2300 proteins, of which, a historical landmark is covering around 90% of the whole expressed proteome of the bacterial genome [12]. One of the most important PTMs, which is studied in great detail and is one of the most important modifications in bacterial pathogenic organisms is phosphorylation [42, 43]. These phosphorylation studies are important because they help in understanding the infection mechanisms of these pathogenic bacteria like *Legionella pneumophila* to cause diseases [44]. Interactomics has fundamental features of connecting the regulatory molecules like kinases, phosphatases, and their target substrates in a way highlighting the interaction between different kinases families phosphorylating each other [45, 46].

The Biological Significance of Post Translation Modifications and their Implications

PTMs are not DNA encoding molecules, but occur during or after protein synthesis, hence are important in cell signaling mechanism [33]. They are fundamental strategies for the adaptation of cells in response to change in the environment as they play important roles in regulating protein function. PTM regulates crucial protein turnover and stability, cell-cell interaction, subcellular localization, differentiation [47]. There are many cellular processes, which increase the events of PTMs like aging, stresses, inflammation, apoptosis *etc.*[48]. The following Table **1** list shows commonly occurring PTMs and their role in biological functions.

Table 1. Common PTMs and their cellular roles.

Target Substrate	PTM	Biological Significance	References
Ser, Thr, Tyr, His	Phosphorylation and dephosphorylation	reversible activation inactivation of enzyme activity, modulating molecular interactions in signaling cascades	[49]
N-terminal and K-residue	Acetylation and deacetylation	modify transcription factors, nuclear import factors, and alpha-tubulin	[50]
Diverse variety of proteins	Ubiquitin	degradation by the proteasome	[51]
Membrane proteins	Sumoylation	nucleocytoplasmic signaling, transport, replication, and regulation of gene expression	[52]

(Table 1) cont.....

Target Substrate	PTM	Biological Significance	References
Lys	Methylation	protein-protein interactions and enhancing protein thermostability	[53]
Ser/Thr	Hydroxylation	cell growth and cycling	[54]
	Phosphorylation	Signal transduction	[55, 56]
Proteins	Acetylation	bacterial chemotaxis, DNA replication, and virulence	[11, 57-59]
Cys, Ser, Thr ,Lys	Acetylation	regulatory role by changing the biochemical characteristics such as their charge, stability, and interactions	[57, 60]
Lipid	*N*-acyl *S*-diacylglyceryl Cysteine	Helps to anchor hydrophilic proteins to membranes, or membrane-associated adhesion, nutrient uptake, cell division, environmental adaptation and in pathogenesis	[60]
Proteins	Glycosylation	central role in modulating protein folding, stability, and signaling	[61]

Nowadays, it is very well reported that PTMs are efficient strategies to modify the infection-related proteins and their intracellular localization. But the first report of modified protein was documented in *Corynebacterium diptheriea* around 40 years ago, where diphtheria toxin ribosylates the elongation factor-2 (EF-2) [62]. These PTMs are fatal to the host cell as they block the translation process and subsequently lead to cell death.

Later on, many bacterial proteins were found to be modified which help to disturb the host defense, and perturb the host cellular pathways for their replication [63, 64]. PTMs not only occur in bacterial proteins but also modify the host proteins for their survival. The host protein modification is brought by effectors (intracellular proteins) that secrete toxins through the plasma membrane, where they are capable of cooperating with the interior host cell proteins *via* secretion system pathways [64].

Acetylation is one of the universal PTMs, which occurs by both processes such as enzymatically (by acetyltransferases utilizing acetyl coA) [65] and non-enzymatically (by metabolites like acetyl phosphate) [66]. Pathogenic bacteria *Porphyromonas gingivalis, a* causative agent of periodontal tissue injury, secrete gingipain proteases, which are the major virulence factors responsible for the downfall of the host immune system. These enzymes are secreted as inactive zymogens, which are converted into active proteases by PTMs. A recent study on the understanding of secretion and activation of these proenzymes has explored the identification of three acetylated lysines in RgpA, RgpB, and Kgp, which suggested that acetylation is a crucial factor for *P.gingivalis* maturation and gingipain activation [67]. Proteins in many pathogenic organisms like *Campylobacter jejuni* [68], *Neisseria gonorrhoeae* [69], and *Burkholderia cepacia* [70], have been glycosylated and suggested the significance of proteins glycosylation in the bacteria.

Methylation also adds a level of complexity to the critical functioning of the proteome. Bacterial pathogens use methylation as the counter reply to evade the host immune system during recurrent infection. The following Table **2** shows the role of methylation in different pathogenic organisms with their survival strategies.

Another remarkable and ubiquitous protein modification is phosphorylation that is regulated by two important enzymes, *i.e*, kinases and phosphatases. It is tightly coordinated and regulated with varying activating inhibiting phosphorylation [76, 77]. Due to the recent advances made in mass spectrometry methods, many novel phosphorylation and acetylation sites have been scaled. These analyses have helped to capture thousands of phosphorylation and acetylation residues and store them in large inventories. Some of the very famous inventories are currently PHOSIDA [78], Phospho ELM, [79] and PhosphoSite [80].

Table 2. Pathogenic organisms having different methylation PTMs and their significance.

Pathogenic Organisms	Type of PTM	Residues /Site of Methylation	Function	References
Enteric bacteria	O-Methylation	Glutamic acid	Helps in Chemotaxis (Che-R is methylated by methyltransferases)	[71]
Mycobacterium tuberculosis	secreted 5-methylcytosine-specific DNA methyltransferase (Rv2966c)	Cytosine	Host gene expression and host-pathogen interaction, host epigenetic machinery to alter transcription	[72]
Rickettsia,	Methylation	outer membrane protein B (OmpB)	Virulence	[73]
Rickettsia,	Trimethylation	outer membrane protein B (OmpB)	Host adhesion, attachment, and invasion,	[73]

(Table 2) cont.....

Pathogenic Organisms	Type of PTM	Residues /Site of Methylation	Function	References
P. aeruginosa	Trimethylation	elongation factor (EF-Tu) on its Lys5 by elongation factor Tu modifying enzyme	Thermostable, bacterial attachment to respiratory cells through platelet activating receptor	[74]
Mycobacterial spp	Methylation (Negative response)	SUV39H1 methylates surface-exposed mycobacterial protein HupB	Decreases bacterial adhesion and survival (negative Methylation response)	[75]

Analogous to the eukaryotic proteasome system where protein substrates are tagged with polyubiquitin tag and directed to degradation pathways, prokaryotes also used prokaryotic ubiquitin-like protein (Pup) [81]. The groundbreaking studies were conducted on *Mycobacterium smegmatis* (Msm) and *Mycobacterium tuberculosis* (Mtb) [82]. Their proteome studies have shown that a small protein called Pup is tagged post-translationally to the various lysine residues to protein substrates which target proteasomal degradation [83].

PTM Stories Unfolded in Some Pathogenic Organisms

When bacteria are introduced to a stressful environment, they undertake diverse physiological adaptations to survive the fluctuating surroundings. Biofilm formation is one of the mechanisms in which an unprecedented multicellular community organization is observed. The biofilm initiation starts when bacteria attach to any surface (or tissues) [84, 85]. *Pseudomonas aeruginosa* (the causative agent of chronic lung infection) is capable of forming biofilm (grey color). Bacteria produce alginates, which is a linear polysaccharide of D-ManA and C5 epimer L-GulA linked *via* β-1, 4-linkages [86]. These are synthesized as PolyM in the cytoplasm and then exported to the periplasm. The major proteins in alginates that undergo PTMs are AlgF, AlgI, AlgJ, and AlgX, which are acetylated (yellow circle in Fig. (4) at O2 or O3 hydroxyl positions of polyM in the periplasm. Bacteria lacking these acetylation form very small, unstructured microcolonies, which suggest a major role of this PTM in virulence, adhesion, and resistance to phagocytosis by the host cell, and ROS scavenger [87 - 89].

Another very important PTM is epimerization (pink diamond in Fig. **4**) undergoing in the proteome of *P. aeruginosa,* which occurs on the AlgG (green circle coding G). These changes are crucial for determining the strength of biofilm. These modifications interact with GluA and calcium ions and make stronger, more viscous, and thick gels in the biofilm, which help them to thrive in the extremities of the host cell's immune system [90, 91]. In addition, *P. aeruginosa* synthesizes the other two important components, which are Pel (PEL) and Psl (PSL) polysaccharides associated with the non-mucoid biofilm formation [92]. PEL is made of many glucose units which are deacetylated (yellow rings) to form biofilms [93].

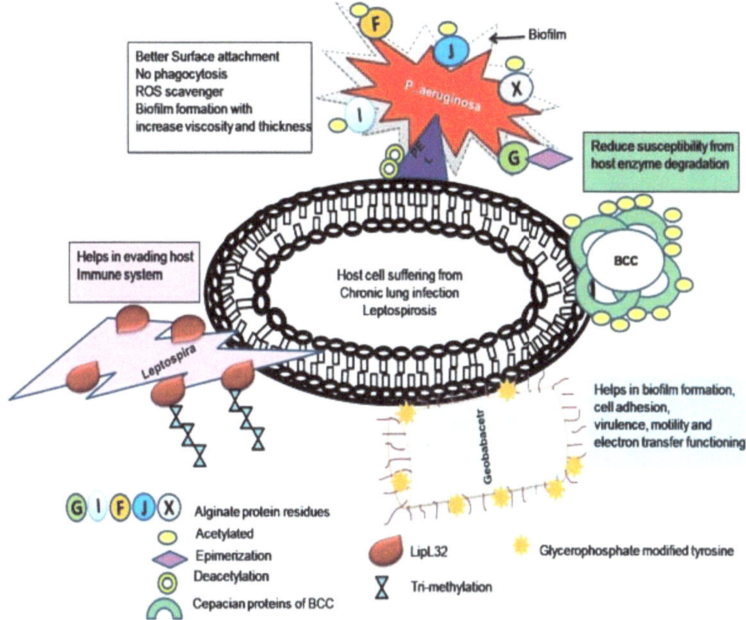

Fig. (4). Some pathogenic organisms showing PTMs and their implications.

BCC is a group of 17 dissimilar bacterial species that are linked with beneficial and harmful effects. The important protein present in these bacteria is cepacian (green semicircle in Fig. **4**). It is acetylated (yellow circles) at 12 different locations with the help of enzymes acetyltransferase BceOSU [94, 95].

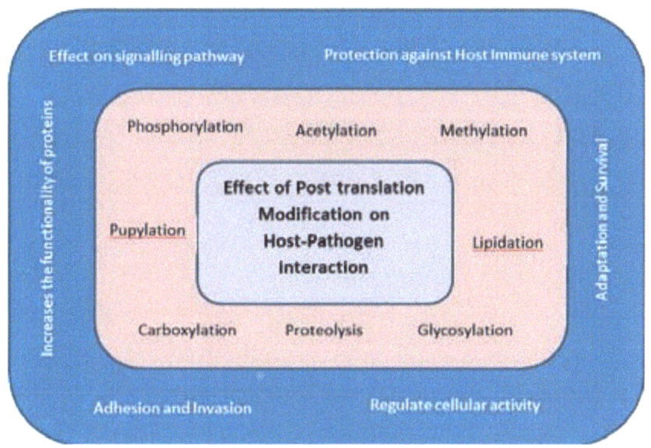

Fig. (5). PTMs involved in host-pathogen interaction and their implication.

Role of PTMs in Host-pathogen Interactions

- Several mechanisms are adopted by the pathogens (like viruses, bacteria, nematodes and fungi) (Fig. **5**) to alter cellular processes during their complex interactions with the host. Such interactions are mediated by pathogen proteins and these proteins act inside host cells to alter cellular processes and encourage infection. These virulence proteins apply diverse mechanisms to control host proteins; one of these is posttranslational modifications (PTMs).

- The role of PTMs is essential in cellular physiology, so pathogens interfere in various ways with the PTMs of their host to promote their survival and replication. Bacterial pathogens interfere with the host PTMs, which is a widely used strategy that allows alteration of host or bacterial key factor activities included in infection [96].

- A variety of bacterial effectors trigger the influence of bacteria on particular host PTMs effectors, which are located either at the bacterial surface or secreted. These effectors may interact with the plasma membrane or intracellular host proteins. The latter case is observed for intracellular bacteria and also for toxins secreted by extracellular bacteria and capable of penetrating inside the host cell, or for effectors directly injected into the host cell through type III or IV secretion systems (T3SS, T4SS). Virulence proteins also use PTMs to manipulate and regulate host proteins, as the host uses PTMs to regulate cell processes. Some virulence proteins are enzymes that can directly mediate target PTMs, while others indirectly mediate host protein PTMs by recruiting host components [31].

- The PTMs (methylation, glycosylation, lipidation, carboxylation, nitrosylation, phosphorylation, acetylation, and oxidation) are not genetically encoded and have shown many different effects through the verities of cellular processes in which it is involved. A specific modification of amino acids in a protein may, in some cases, alter, expand, or enable its enzymatic activity. They can also have a specific effect on protein conformation or a protein surface charge. Modifications of proteins, such as phosphorylation, methylation, or acetylation, may directly or indirectly alter or facilitate interactions with proteins or other cellular components. Specific hydrophobic changes such as lipidation may target the modified membrane proteins. PTMs such as glycosylation and bacillithiol supplementation can also protect proteins and extend their activity [97, 98].

- Fascinatingly, host proteins seem preferred targets for PTMs that are caused by bacteria. Rho GTPases are attacked by a broad variety of modifications, resulting in either their constitutive activation (through deamidation or polyamination)/inactivation (through AMPylation, ADP- ribosylation, glucosylation, and proteolysis), or degradation (through polyubiquitylation). Rho GTPases are the key molecular switches in eukaryotic cells. The targeting of these regulators may therefore be an effective strategy developed by pathogens to alter the behavior of several different proteins in a given pathway, thus inducing a coordinated response of the host cell by targeting only one element. Targeting Rho GTPases also represents the well-established crucial function of remodeling host cell cytoskeleton in many aspects of bacterial infection. Consistently, several other cell cytoskeleton components have been reported to have post-translational modifications by bacterial pathogens. However, pathogen targeting of Rho GTPases is likely to go beyond cytoskeleton modification as these proteins are involved particular in an innate and adaptive immune response.

- ***Protein phosphorylation***, which serves as a reversible molecular switch, provides a mechanism for regulating protein function in nearly all cellular processes and is important for the rapid response of cells to internal and external signals. The covalent binding of adenosine triphosphate (ATP) phosphate groups to serine, threonine, and tyrosine residues are accomplished by protein kinases and the reverse reaction by protein phosphatases [99].

- Protein function is regulated by phosphorylation, which can either directly modulate the enzymatic activity or provide a docking site for interactions between intra or intermolecular proteins. Such changes affect the subcellular position and protein turnover, as well as the interplay between phosphorylation and other posttranslational protein modifications (PTMs). In this way, Kinases and their substrates form dynamic complexes and temporary information processing networks that promote cell-cell communication and cellular responses to changing environmental conditions [100].

- Phosphorylation-based signaling leads to innate and adaptive immunity in important ways, for example in the protection of host cells against pathogenic bacteria. Main immune system processes, such as differentiation, development of cytokines/chemokines, inflammation, and bacterial killing, are largely mediated by protein phosphorylation and, subsequently, by a corresponding protein kinase. In the innate immune system, microbe-associated molecular patterns are recognized by unique pattern recognition receptors (PRRs) that cause pro-inflammatory and antimicrobial reactions when engaged. Such receptors are also an important link to the adaptive immune system that eventually results in infection resolution and immunological memory. For example, the Toll-like receptor (TLR) family of PRRs has been extensively studied. Stimulation of TLR proteins results in the subsequent activation of several kinases, including interleukin1-associated receptor kinases (IRAK1, 2, and 4), mitogen-protein kinases (MAPKs; *e.g.* MAP3K7/TAK1, p38 alpha, and JNK) and the subsequent phosphorylation of downstream targets, such as activator protein 1 (AP1) and nuclear factor κB (Nfκb), that act as master transcriptional regulators during the induction of proinflammatory and anti-apoptotic mediators. Inflammatory disorders, autoimmunity and pathogenesis are correlated with deregulation of these signaling systems [101].

- On the other hand, it has recently become evident that host signaling pathways associated with main processes, such as membrane and cytoskeleton dynamics, autophagy, vesicle transport, cell death, inflammation, and immunity, are manipulated during an infection. Such bacterial pathogens have developed mechanisms, such as the development of unique toxins, effector molecules, and virulence factors, by which they subvert and regulate such signaling pathways to their advantage through promoting bacterial adherence, survival, reproduction, and spread. Host signaling interference with kinase-mediated phosphorylation is a valuable technique used by many pathogens.

- One of the most common post-translation modifications (PTMs) in all kingdoms of life is **protein glycosylation**, that is, enzyme-catalyzed covalent attachment of glycans to the amino acid side chains in proteins. A large number of proteins are glycosylated in various pathogenic bacteria, including *Pseudomonas aeruginosa, Neisseria meningitides*, Haemophilus influenzae, Campylobacter jejuni, *Mycobacterium tuberculosis, Streptococcus parasanguis*, diffusely adhering *Escherichia coli* [102 - 107].

- It is also important to note that certain bacterial toxins or virulence factors bear glycosyltransferase activities designed to alter their host targets during infection. Protein glycosylation affects all protein functions-their structure, behavior, interactions with other molecules, cell or organism half-life. Immune recognition often depends on the immune cells and receptors communicating with glycosylated molecules. A wide variety of potential glycan structures and linkages improves the versatility of proteins [61].

- Among bacteria, glycosylation creates a much more complex range of glycoconjugates that are often species-or strain-specific. Most bacterial glycoconjugates are an important part of the bacterial cell wall and contribute to the structural integrity of the bacterial cells. Besides, bacterial glycosylated cell surface structures mediate adhesion and contact with the environment or host. While the structural features of bacterial surface glycans have been well defined many of them, including those in pathogenic bacteria, remain unexplored. In general, pathogenic bacteria use glycosylation for two reasons; they synthesize host-like glycan structures to hide from the host immune system and, conversely, create glycosylated proteins that can bind host immune molecules more effectively and thus influence their activity [108].

- In many bacteria, protein glycosylation systems are also associated with essential biological processes such as pathogenicity, immune evasion, and host-pathogen interactions, which suggest the importance of protein-glycan bonding. In the same way, host protein glycosylation has been involved in the antimicrobial activity as well as in promoting the growth of beneficial strains. Also, few pathogens modulate host glycosylation machines to promote their survival. Protein glycosylation, as present in all kingdoms of life, is one of the most common post-translation modifications (PTM) of proteins. This consists of the covalent binding of glycans to side chains of amino acids; this reaction is catalyzed by an enzyme. Glycosylation has important impacts on eukaryotic cells on protein folding, conformation, distribution, stability, and behavior. Specifically, the glycoprotein sugar chains are essential to maintain the order of intercellular interactions among all differentiated cells in multicellular organisms [109].

- Bacteria are capable of producing extraordinarily large amounts of specific and complex glycans, primarily attached to the cell surface, and secreted molecules. These glycoconjugates can be used by bacteria as a collection of unique and common ligands that interact directly with the host. Early studies suggest that glycosyltransferases present in bacteria mainly alter their surface proteins in streptococci and staphylococci, such as HMW1/2 in NTHi and SRRPs. The new BAHT family has expanded the list of targets for glycosylation to a subfamily of carriers, including prototypical AIDA-I and TibA in enterobacterial pathogens [76, 77]. Glycosylated bacterial surface proteins and auto transporters, irrespective of whether they are mono-or polyglycosylated, frequently serve as adherence factors for bacterial attachment to the host cell, given the diverse pathophysiologies of the bacteria involved. Not only do bacterial glycosyltransferases participate in this initial infection phase, but they also modulate host responses through direct modification of host proteins. The classic example is the large clostridial glucosylated toxins that modify Rho GTPases to manipulate the dynamics of host actin cytoskeletons [110, 111].

- In many bacteria, protein glycosylation systems are also associated with essential biological processes such as pathogenicity, immune evasion, and host-interactions, which suggest the importance of protein-bonding. Similarly, host protein glycosylation has been involved in both antimicrobial activities and the promotion of beneficial strain production. Besides, few pathogens significantly modulate machinery for host glycosylation to facilitate their survival.

- ***Protein acetylation and succinylation:*** Recent global studies of lysine acetylation revealed that this reversible and evolutionarily conserved protein modification is ubiquitous in bacteria. Acetate can be enzymatically added to and removed from lysine side chains by lysine acetyltransferases and deacetylases, respectively. These PTMs affect protein structure, protein-protein interaction, DNA–protein interactions, and cellular localization and can be used by bacteria to rapidly adapt to environmental changes [112].

- ***Protein pupylation (ubiquitin-like modifications):*** PTMs by covalent attachment of other small proteins are a prominent feature of eukaryotes where ubiquitin or similar small protein modifiers prevail. Actinobacteria (and sporadic members of a few other lineages) encode a small altering protein functionally similar to ubiquitin called' procaryotic ubiquitin-like protein' (Pup). Pup is an intrinsically disordered small protein (60-70 residues) that attached to the lysine side chains of target proteins through the side chain carboxylate of its carboxy-terminal glutamate residue known as pupylation. Under stress conditions, pupylation and proteasomal degradation play a role in survival. Whether pupylation directly controls the activity of certain targets, remains to be identified [113].

- ***Protein lipidation***: Protein lipidation is a reversible enzymatic PTM that involves the attachment of lipid chains to the residue of cysteine through a thioester bond. Lipidation stimulates secreted bacterial toxins and leads to the adhesion and invasion of host cells and defense against the host immune system [114].

PTM Role in Drug Resistance

PTM is also very useful for antibiotic resistance, and it gives bacteria an additional sheath to provide resistance from antibiotics. Many antibiotics available in the clinic today directly inhibit bacterial translation. Despite the past success of such drugs, their efficacy is diminishing with the spread of antibiotic resistance. Through the use of ribosomal modifications, ribosomal protection proteins, translation elongation factors, and mistranslation, many pathogens can establish resistance to common therapeutics.

Drug resistance in bacteria may be of two types: intrinsic or acquired. Intrinsic resistance, resistant bacteria mutants exist before drug/antibiotic treatment and in adaptive resistance where bacterial pathogens develop resistance by altering their molecular physiology in response to the antibiotic encounter (Fig. **6**). Molecular experiments indicate that, like eukaryotes, prokaryotes also exhibit a significant number of post-translational changes (PTMs) in their proteins. Bacteria adopt a variety of mechanisms to prevent the effects of medications. Bacteria are small unicellular organisms which have small genomes to increase functional versatility, and use post-translation modifications that add chemicals and provide additional functionality.

Resistant bacterial strains have also evolved alternative ways of controlling metabolism by manipulating PTMs that help to establish drug resistance through various mechanisms. There are many known drug resistance mechanisms in bacteria that include: change in cell wall properties, active efflux pumps, drug target site alteration, DNA repair system, alternate metabolism pathways.

This chapter lists examples of how PTM modulation can induce drug resistance in bacteria. Throughout all the mechanisms addressed for resistance to bacterial drugs, we usually address basic chemical changes to either proteins or other bacterial cell components (DNA, lipopolysaccharide, *etc.*) or sometimes even to the drug itself that can help bacteria develop resistance [115]. Although all of these chemical modifications are not strictly PTMs in the classical sense, we mainly focus on PTM level modulations but also address few non-PTM adjustments to ensure wide coverage of the resistance problem. Knowledge of PTMs role in drug resistance will enable us to design new therapeutic treatments for bacterial diseases and effectively address emerging resistance problems.

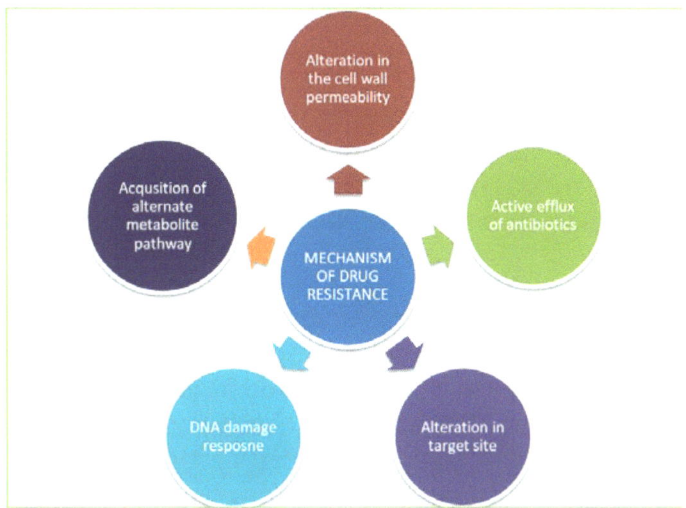

Fig. (6). Different mechanisms of drug resistance in pathogenic organisms.

Mechanisms of Bacterial Drug Resistance

The cell wall, being the most robust outer layer of bacterial cells, acts as the primary mechanism of defense against antibiotic care. The ultrastructure of the cell wall enables selective nutrient permeability and reduced bacterial protection mechanism for the drug as the first line of defense. Modification in the cell wall that leads to antibiotic resistance in bacteria is as follows:

1. By increasing the thickness of the outer membrane.
2. By increasing cross-linking of cell wall.
3. By decreasing the size of porins.
4. By decreasing the negative charge of cell wall.

Protein's complex interaction with sugar and lipid decorations plays a key role in their sensitivity towards antibiotics and resistance growth. Antibiotics achieve entry into the outer membrane by diffusion into porin proteins or by lipid chain-mediated translocation. Another potential entry mode is through permeabilization of the outer membranes. To combat these, bacteria have built mechanisms to block their entry. Pathogens have considerable resistance to antibiotic entry, as demonstrated by changes in the outer membrane structure, either by thickening it or by adding components that increase the cross-linking of outer membrane components to create a compact packing. In drug-resistant species, the pore size was found to be smaller, reducing the entry of antibiotics.

Ribosome alteration is one of the main antibiotic-resistance methods. Bacteria use methyltransferase to methylate either the large 23S subunit of rRNA or the small 16S rRNA. Through this methylation, antibiotics binding to these rRNA prevented, which inhibited drug activity. Erythromycin causes a ribosomal pause on the leader peptide, for this methylase responsible for the 23 s rRNA ermC dimethylation. This Pusing causes the transcript to form a structure in which the shine dalgnaro sequence for ermC is released, allowing the ermC to be translated. ErmC then methylates the nascent peptide exit tunnel and prevents the binding of erythromycin [116].

In the case of an organism that maintains numerous copies of rRNA genes, rRNA methylation is an ideal mode of resistance. Since ribosomes can be changed without the need to acquire the same gene mutation in the genome. The microbes acquire drug resistance through PTM, which by decreasing the pore size decreases the permeability of the cell wall and PTM in efflux pumps. In addition, drug resistance is often obtained by deliberately amending antibiotics. It induces target site modification and provides a reaction to DNA damage. This also assists in obtaining alternate metabolite pathways. Each of these effects gives microbe's drug resistance.

Different PTMs Present in Bacteria

Protein methylation is also known as alkylation, where methyl is the alkyl group added on amino acids like Lys, Arg, Gln, Glu with the help of methyltransferases. Bacterial protein may get methylated and they can do methylation of human protein. The methylating property of bacterial toxin helps bacteria for infection. Toxins play an important role in the pathogenesis of Bacillus anthracis by subverting the host defense system. Here bacterial protein is methylating human histone protein. Bacterial toxin protein, BaSET (that is suppressor-of-variegation, enhancer-of-zeste, trithorax protein) methylates 8 lysine residues in human histone H1 protein in macrophages, resulting in repression of NF-kappaB activity, so increasing its virulence (Table **3**) [117].

Table 3. Different types of PTM found in microorganisms.

Type of PTM (Mass Change)	Amino Acid Modified	Its Role	Organisms Exhibiting this Modification	Ref.
Methylation- methyl groups (14 Da): N methylation	Lys, Arg	Affects virulence property of bacteria, modification of nucleoid-associated proteins (NAPs), which can activate or inhibit particular gene	*L.interrogans,* Rickettsia, *Bacillus anthracis, E coli, P. aeruginosa, Sulfolobus islandicus*	[73, 117, 118, 122, 124]
Methylation- methyl groups (14 Da): O Methylation	Gln, Glu	Methylation of methyl-accepting chemotaxis proteins (MCPs), which play an important role in Chemotaxis according to the external environment, methylation contributes to immune evasion during infection by developing potential antigenic epitopes	*E. coli* *L. interrogans*	[127, 128]
Acetylation- acetyl groups (42 Da)	Ser, Thr, Lys, protein amino termini	Affect cellular processes of carbon metabolism, helps in adaption against environmental changes, acid stress, high temperature and reactive oxygen species, increase virulence property	*M. Tuberculosis* *E. coli* *V. cholera Yersinia pestis* *Salmonella enteric*	[130, 131, 134, 139] [57, 135, 137]
Phosphorylation- phosphate groups (80 Da)	Ser, Thr, Tyr, His, Asp, Arg, Cys	Role in bacterial stress response, the developmental process in bacteria like spore germination by affecting transcription and translation. can help in osmotic stress , increasing pathogenicity, can help in decreasing time for treatment of tuberculosis, will regulate diverse cellular processes, affect host defense system	*Bacillus subtilis, Staphylococcus aureus.* *Escherichia coli* and *Salmonella Streptococcus agalactiae Mycobacterium tuberculosis Yersinia enterocolitica*	[140 - 142] [143] [144] [145, 146] [147]
Hydroxylation		Affects cell cycling and growth by changes in ribosomal protein in translation	*Escherichia coli*	[54, 148]
Lipidation (variable)	Cys,Ser, Thr	Helps in increasing virulence by affecting immune cells, colonization and pathogenesis, resistance to antibiotic treatments, pathogen to escape the phagosome and survive in macrophages, enhance its antibacterial activity against other bacteria	*Brucellaceae Streptococcus agalactiae Neisseria meningitides Francisella tularensis, Bacillus subtilis*	[149 - 152] [153]

(Table 3) cont.....

Type of PTM (Mass Change)	Amino Acid Modified	Its Role	Organisms Exhibiting this Modification	Ref.
AMPylation	Tyr, Ser, or Thr	Responds to unfolded protein and endoplsmic reticulum stress	*Enterococcus faecalis*	[154]
ADP-Ribosylation	Arg	Affects virulence property of opportunistic pathogen	*Pseudomonas aeruginosa*	[155]
Glycosylation-oligosaccharide structures (2–3 kDa)	Asn, Arg, Ser, Thr, Cys	Affect pathogenicity, rescue stalled ribosome synthesis, aid colonization of bacteria in the gut	*Campylobacter jejuni, Shewanella oneidensis, P. aeruginosa, Lacto bacilli*	[156 - 158] [28]
Carboxylation (43Da)		Activates the antibiotic (β-lactam) sensor, helps in DNA damage repair	*Acinetobacter baumannii, Thermus thermophiles, E. coli*	[159] [160]
Nitrosylation (28Da)	Cys	Transcriptional regulator OxyR, results in decreasing DNA binding, so affecting the transcription process	*E. coli P. aeruginosa*	[161]
Protein pupylation and ubiquitin- like modifications		Help in sporulation, regulation of bacterial cell metabolism, help in amino acid recycling in nutritional starvation, help in recovery from the DNA damage control	*Streptomyces coelicolor, C. difficile, M. smegmatis, M. tuberculosis*	[162 - 166]

Like eukaryotic histones, nucleoid-associated proteins (NAPs) are present in bacteria, and the four most abundantly present NAPs in *Escherichia coli* (H-NS, HU, IHF, and FIS) were observed by proteome mass spectrometry [118]. Protein methylation in microbes is involved in chemotaxis, signal transduction, energy metabolism, and survival under high temperatures by stabilizing protein. Methylation would be important for microbial survival under the extreme environment of the hydrothermal vent [119]. Methylation of protein is involved in protein-protein interactions, and that enhanced protein thermostability [53]. In *Agrobacterium tumefaciens*, electron transfer flavoprotein (ETFß) transfers an electron from various dehydrogenases, once methylation is done on ETFß it can no longer transfer the electrons [120]. In the hyperthermophilic archaea, *Sulfolobus islandicus*, lysine methylation occur at high temperatures (over 70 °C). The helicase activity of mini-chromosome maintenance (MCM) increased for survival at high temperature [121].

Methylation is sub-divided into two: i) N-methylation and ii) O-methylation. N-methylation includes the transfer of a varying number of methyl groups onto the terminal amine of Lysine or Arginine amino acid. Lys can accept up to 3 methyl groups and Arg can accept up to 2 methyl groups. In *L.interrogans* 64 Lys/Arg methylated sites from 58 proteins were identified. Among 64 Lys/Arg-methylated sites, there were 13 monomethyl, 14 dimethyl, and 13 trimethyl sites of Lys, along with 10 monomethyl and 14 dimethyl sites of Arg [122]. In Rickettsia, two OmpB-specific methyltransferases were observed, RP027-28 catalyzed trimethylation and RP0789 catalyzed mono-, di- and trimethylation [123]. Trimethyl lysine PTM of elongation factor Tu (EF-Tu) is carried out by methyltransferase EftM, resulting in a structure that mimics phosphorylcholine present in platelet-activating factor (PAF). This helped bacteria to attach the platelet-activating factor receptor (PAFR) of respiratory cells and resulted in various respiratory infections. Lys 5 trimethylation of EF-Tu which is found to be surface-exposed, allowed the pathogen to bind to PAFR and helped *P. aeruginosa* to infect. This can help us develop novel drugs to prevent *P. aeruginosa* pneumonia, which is useful for the emergence of multidrug-resistant strains [124].

O-methylation includes the formation of a methyl ester between the carboxylate side chains of Glu. O-methylation plays an important role in the signal transduction pathway of bacterial chemotaxis. In response to chemosensory signals from metabolism and growth state, the bacterial flagellar rotary motor protein is manipulated by effector proteins, like CheY, FRD, YcgR, H-NS, and EpsE [125]. To respond against the changing environment, bacteria move. Bacterial chemotaxis is the movement towards environments that contain higher concentrations of beneficial chemicals and away from the lower concentrations of toxic chemicals. The pathway is composed of chemoreceptors, the histidine protein kinase chemotaxis protein (sensor kinase CheA), and two diffusable response regulators (CheY and CheB). CheY controls flagellar motor switching, whereas CheB controls chemoreceptor adaptation. Chemotaxis plays a role in pathogenicity, stability, symbiosis, and biofilm formation, which help for bacterial survival in a changing environment [126].

Methylation of the methyl-accepting chemotaxis proteins (MCPs) exists in clusters at the outer membrane. *E. coli* cells are also able to respond to changes in response to environmental chemo-effector concentrations after binding to the amino-terminal periplasmic domain of the MCP. Attractants promote counterclockwise rotation of the flagella, resulting in smooth swimming, whereas the repellents promote clockwise rotation, resulting in tumbling. These responses are mediated by membrane-bound methyl-accepting chemotaxis proteins (MCPs) [127]. In *L. interrogans,* Glx-methylation (Glx is either Glu or Gln deamidated by CheB, a bifunctional deaminase) at 11 sites within proteoforms of LipL32. *In-silico* analysis revealed that regions containing Glx-methylation were associated with potential antigenic epitopes, suggesting that methylation may contribute to immune evasion during infection [128].

Acetylation

Acetylation is the addition of acetyl group to amino acids like Ser, Thr, and Lys at the N-terminal of protein. N-acetylation is the addition of either the ε-amines of Lys side chains (Nε-acetylation) or α-amines of protein N-termini (Nα-acetylation). Nε-acetylation can be done by two mechanisms in terms of donation of the acetyl group to the ε-amino group of deprotonated lysine: one is enzymatic, which depends on a lysine acetyltransferase (KAT) and acetyl-coenzyme A (acCoA). And the other is non-enzymatic and depends on the reactivity of acetyl phosphate(acP) [129]. Deacetylases remove acetyl group (NAD+-dependant sirtuins; for example, CobB and Zn+-dependent deacetylases). Various PTMs like KATE are from *M. Tuberculosis* [130], and the deacetylase YcgC from *E. coli* [131]affect protein structure, interactions between two proteins or between DNA and protein, and cellular localization. It helps bacteria for adaption to environmental changes [112].

Lysine acetyltransferases (KATs) can be of two types: GNAT and YopJ effectors. Prokaryotic GNAT-KATs, can be classified into three classes. Their subdivision in types (I to V) is done based on the sequence, arrangement, and type of GNAT domain. GNATs, like YfiQ is a conserved acetyltransferase [132] that is found across many bacterial species. It affects central metabolism in *E. coli* [133], affects the virulence property of *Vibrio cholera* [134]. Along with YfiQ, CobB is involved in the virulence and stress response of *Yersinia pestis* [135]. YfiQ helps for protection against acid stress in *Salmonella enterica* serovar Typhimurium [136], it also protects against high temperature and reactive oxygen species [137]. In *E. coli*, the sirtuin CobB does deacetylation and affect translation and protein function [138].

Acetylation is mainly a key regulator for metabolic enzyme modification involved in many cellular processes such as glucose and gluconeogenesis. Lysine acetylation, which is conserved throughout evolution, removes the charge and makes changes in the shape of the enzyme's site, which is required for substrate or cofactor binding. This will result in switching off enzyme activity of many glycolytic and tricarboxylic acid (TCA) cycle enzymes. So acetylation plays an important role in protein activity regulation [139], *cob* and *yfiQ* genes in *V. Cholera* helps in the metabolism of acetate, which affects its virulence property by various physiological roles in a diversity of host organisms [134].

Phosphorylation

Protein phosphorylation is catalyzed by kinases using the γ- phosphate of ATP as a donor, and the reverse reaction, dephosphorylation is catalyzed by phosphatases. Phosphorylation and dephosphorylation of the specific amino acid are responsible for the activation or deactivation of protein. Phosphorylation of protein plays an essential role in signal transduction to give a response against external or internal stimuli. Various amino acids like arginine, serine, threonine, tyrosine, histidine, and aspartic acid are modified. There are four types of protein phosphorylation: Arg kinases, Two-component systems, Hanks-type kinases, and BY kinases (Table **4**).

Table 4. Important enzymes in PTMs mechanisms.

Amino Acid Involved in PTM	Name of PTM	Enzyme Responsible for PTM
Aspartic acid	Phosphorylation	Two component system
Asparagine	Glycosylation	N-glycosyltransferases
Arginine	Phosphorylation, Glycosylation	Arg kinase, N-glycosyltransferases
Cysteine	Phosphorylation, Lipidation, Thiolation	Phospho transferase, Diacylglyceryltransferases,Thiol peptide
Histidine	Phosphorylation	Phospho transferase, Two component system
Lysine	Acetylation, Acylation, Pupylation	Acetyl transferase, acyl transferase,pup ligase
Methionine	Oxidation	Methionine sulphoxideredustase
Serine	Phosphorylation, Glycosylation	Hanks-type kinases, O-glycosyltransferase

Arginine phosphorylation plays an important role in bacterial stress response in *Bacillus subtilis* and *Staphylococcus aureus*. Arginine phosphorylation in protein acts as a degradation tag so they are targeted to ClpC-ClpP protease, similar to the function of the ubiquitin-proteasome system in eukaryotes [140,141]. For stress response and pathogenicity, McsB acts in two ways: as degradation labeler, it results in degradation of regulatory proteins as well as acts as a transcriptional regulator [167]. YwlE arginine (Arg) phosphatase, by dephosphorylation, plays a major role in spore germination in *B. subtilis by affecting* housekeeping σ factor A (SigA), which results in reinitiation of *transcription and affects* ribosome-associated chaperone Tig to start *translation* [142].

A two-component system of histidine and aspartate kinase aids in 'fast switching' events in bacteria from the environmental stimulus. In response to the stimulus, membrane-bound sensory kinase undergoes autophosphorylation on histidine and activates aspartate, which is a response regulator. Phosphorylated aspartate is involved in various cellular processes like transcription by binding to the promoter regions of target genes [168].

PhoQ/PhoPTCS known to be dedicated to phosphate sensing in *Escherichia coli* and *Salmonella,* which can help in conditions like osmotic stress [143]. In pathogenic bacteria, some TCS sense the presence of hosts and trigger virulence mechanisms. In the case of *Streptococcus agalactiae*, the transcriptional regulator LtdR responsible for bacterial-host interactions and promote persistence and disease progression [144]. Various mechanisms used to escape antibiotic treatment in *Mycobacterium tuberculosis* are the part of TCS, where phosphate-sensing signal transduction system contributes to antibiotic tolerance and is considered as a potential target for the development of novel therapeutics that may shorten the duration of tuberculosis treatment [145]. Bacteria depend on TCS systems for responding to environmental changes. TCS are important for pathogenicity and can be potential drug targets as inhibition of bacterial signaling machinery by inhibiting histidine kinases [169].

Other than histidine kinase, serine/threonine kinases can also phosphorylate TCS response regulators, for example, PknB a serine/threonine kinases from *Mycobacterium tuberculosis* was phosphorylating the TCS response regulator DevR from *Mycobacterium smegmatis* and affect its binding to target DNA [170]. Bacterial Hanks-type kinases are generally transmembrane proteins, containing an extracellular PASTA and intracellular protein kinase domain. The kinase domains trans-autophosphorylate and result in the opening of the active site region, which further phosphorylates various cellular target proteins at serine and threonine amino acids of the protein. BY kinases phosphorylate the proteins at a tyrosine residue. Sirtuins are NAD^+-dependent deacetylases, mDAC, is phosphorylated by the Hanks-type kinase PknA, which regulate diverse cellular processes in *M. Tuberculosis* [146]. Hanks- ype kinases and BY kinases phosphorylate multiple proteins. Several enzymes of pathogens enter into host cells to interfere with host signal transduction pathways [171]. In *Yersinia enterocolitica* Hanks- type serine/threonine kinase YopO, modulate the actin and hampers macrophage phagocytosis [147].

Hydroxylation

In the *E. coli*, ycfD is growth-regulating 2-oxoglutarate oxygenase, which catalyzes hydroxylation on arginine residues in the ribosomal protein Rpl16. Using these new therapeutic possibilities *via* oxygenase inhibition can be found, which affect cell cycling and growth [54]. Ribosomal oxygenases (ROX) affect bulk protein translation and are linked to cell growth. Hydroxylation of proline residues in collagen is catalyzed by procollagen prolyl 3- and 4-hydroxyl. Defects in all three types of collagen hydroxylations have been linked to several diseases. *E. coli* translational process regulated by YcfD (a Fe^{2+}-dependent oxygenase), which is responsible for hydroxylation of the bacterial ribosome. In response to the environment changes, bacteria undergo hydroxylation, which makes changes in metabolism and cell response in hypoxic conditions [148].

Lipidation

Lipidation is a reversible enzymatic PTM that includes the addition of lipid chains to a cysteine residue *via* a thioester linkage by diacyl glyceryl transferase. Lipidation is a prerequisite for the anchoring and export of proteins, which is crucial for their function. *S*-palmitoylation affects the association of the protein with membranes, stability, and trafficking. The role of the palmitoyl moiety is associated with infections because it is upsetting the recognition of bacteria by innate immune receptors [149].

In *Streptococcus agalactiae*, lipidation helps in colonization and pathogenesis [114]. *Brucellaceae* are a group of pathogenic intracellular bacteria known to use host enzymes to palmitoylate bacterial proteins PrpA affect B-cell activity, result in chronic infection by manipulating immune responses of the host [150]. *Clostridium difficile*, which causes gastrointestinal disease, lipidated proteins in bacteria aid cellular processes like sporulation and pathogenesis [172]. Protein lipidation aids bacterial virulence, for example, by promoting attachment and avoiding recognition by innate immune receptors [149]. Lipidation in *Neisseria meningitides* enhance resistance to antibiotic treatments (rifampicin and ciprofloxacin) [151]. Palmitoyl lipoprotein help *Francisella tularensis* to escape the phagosome and replicate within the cytosol of infected macrophages [152]. Lipidation of the synthetic peptide in *Bacillus subtilis* enhance its antibacterial activity against both gram-positive and negative bacteria [153].

AMPylation

FIC proteins catalyze PTM by the addition of AMP called AMPylation or adenylation. FIC proteins are bifunctional, can also do de-AMPylation of BiP/GRP78, a key chaperone of the unfolded protein response and endoplasmic reticulum stress. The functionality of FIC depends on glutamate and Mg^{2+} and Ca^{2+} concentration in *Enterococcus faecalis*. Here BiP is a key component of the unfolded protein response (UPR). If UPR is present for a too long time it has shown damaging effects and can cause disease, therefore it serves as a desirable target for drug development. Other examples of FIC in pathogenic bacteria are Neisseria FIC, BartonellaVbhT/VbhA and Clostridium FIC [154].

ADP-Ribosylation

In *Pseudomonas aeruginosa*, virulence is mediated by type III secretion (TTS) of bacterial proteins. Exoenzyme S (ExoS) is the bifunctional enzyme responsible for ADP-ribosyltransferase (ADPRT) activities on Ras superfamily and ERM family proteins. These proteins mediate signal transduction, and their modification so they can produce cellular changes that will facilitate the infectious process and interrupt multiple host cell processes [155].

Glycosylation

Sugar is attached in the glycosylation PTM and based on the attachment of sugars; glycosylation is divided into three types; N- linked (sugar attachment on the nitrogen atom of asparagine or arginine), O- linked (sugar attachment on hydroxy groups of mainly serine or threonine), and S- linked (sugar attachment on thiol group of cysteines). In *Campylobacter jejuni,* N- glycosylation of proteins affects the virulence property [156]. Most of the glycosylated bacterial proteins are present on the cell surface in *Shewanel laoneidensis,* and *P. aeruginosa* translational elongation factor P (EF- P) is activated by arginine rhamnosylation [157]. Rhamnosylated EF- P rescue stalled ribosome synthesis [158] and result in ribosome generation and affect the translation. Glycosylation affects interactions of bacteria like lactobacilli and the lining of the human gut [28].

Glycosylation of cell surface bacterial proteins enhance the adhesion of bacteria to the host and is crucial for the colonization in the gastrointestinal tract [158]. Colonization of good bacteria helps the human host by increasing good microflora in the gut, while colonization of bad bacteria helps pathogenic bacteria to increase the pathogenicity. Therefore glycosylation affects a broad range of transmembrane and secreted proteins, which are involved in cell-cell interaction and virulence.

Carboxylation

In *Acinetobacter baumannii,* carboxylation and decarboxylation of the β-lactamase enzyme were observed in Lys amino acids. Carboxylation activates the antibiotic (β-lactam) sensor protein BlaR1 from *S. aureus* [159]. Carboxylation of UV damage endonuclease in *Thermus thermophilus* suggested that it is required for proper catalysis, and preventing the increased incision of undamaged DNA. Therefore it helps in DNA repair and maintains the integrity of bacteria [160].

Nitrosylation

This PTM is derived from condensation with reactive nitrogen species. S-nitrosylation appears to convey endogenous nitrosative stress during anaerobic metabolism, which is analogous to the oxidative stress entailed by oxidative (aerobic) metabolism. S-nitrosylated protein in *E. coli* and *P. aeruginosa* make changes in transcriptional regulator OxyR, result in a decrease in DNA binding and affect the transcription process. The role of this PTM in cell signaling and the immune response is yet to be discovered.

Protein Pupylation and ubiquitin- like Modifications

Pupylation is the covalent attachment of the 'protein ubiquitin-like protein' (Pup) result in the degradation of tagged protein, which was reported in *M. Tuberculosis*. Pupylation is important for sporulation in *Streptomyces coelicolor* [162] and as a regulator of spore germination in *C. difficile* [163]. Pupylation is also involved in the regulation of bacterial cell metabolism. Pupylation helps amino acid recycling in nutritional starvation [164]. In *M. smegmatis,* it helps in the survival under nitrogen starvation [165] by nitrogen assimilation [173]. In mycobacterium, pupylation could also contribute to drug resistance [166].

Proteomic Approaches Used in Exploration of Microbial PTMs

The bacterial PTM is necessary for the survival of bacteria in adverse conditions and environments. The major importance of PTM is:

1. Role in drug resistance.
2. Host-parasite association.
3. Interaction with other proteins,
4. Increase in pathogenicity and defense mechanisms.

In this section, we have discussed the techniques and tools for identifying the presence of PTM.

The involvement of PTM in biological pathways as well as in metabolic activity is an important part of the study in biomedical research. It is also important how cells modify themselves by the use of the PTM strategy for increasing its fitness in unfavorable environments.

Global and quantitative analysis of pathogen proteome assists in the apprehension of the events at the molecular level that are regulated with the onset of infection. At the proteome level, the changes (dynamic) are usually studied by Shotgun proteomic approaches. This technique is usually employed for bacterial proteome and their virulence pathways. This technique involves the extraction of proteins from the organism, followed by digestion with respective endoproteases and analysis by LC-MS technique [174]. The other alternative approach is top-down proteomics that involves the characterization of bacterial proteomes by utilizing the intact proteins and is also employed to study post-translational modifications, which occur during infection in bacterial proteome and host proteome [175].

Proteomic studies involving bacteria are also used for the detection of biomarkers, which act as a target for drug-based theory [176]. Denovo sequencing is utilized when the DNA sequence of an organism is incomplete, FACS (Fluorescence

Activated Cell Sorting) are generally utilized for isolation and separation of bacterial cells from host cells. A study [177] conducted on Epsilon toxin (ETX) induced by *Clostridium perfringes*, occurs in a stepwise and ordered fashion by intestinal contents and thereby resulted in three different ETX speciation (27 KDa). Proteases such as serine, trypsin, chymotrypsin, carboxypeptidases are required in the activation of a toxin.

The other proteomic tools used, such as 2D-PAGE and 2D-DIGE exploit only one single gel that estimates the differences between two protein samples. SILAC (Stable Isotope Labeling by Amino acid in Cell Culture) integrates labeled isotopes and utilizing one or two amino acids of the growth medium. Furthermore, ICPL (Isotope coded Protein Label) involves the labeling of amino groups which used in Mass Spectrometry [178]. These techniques offer several advantages and disadvantages, which are discussed briefly in the Table **5** below [179].

The following flow chart discusses the sequentially important steps involved in proteome analyses, sample preparation and tools to study PTMs.

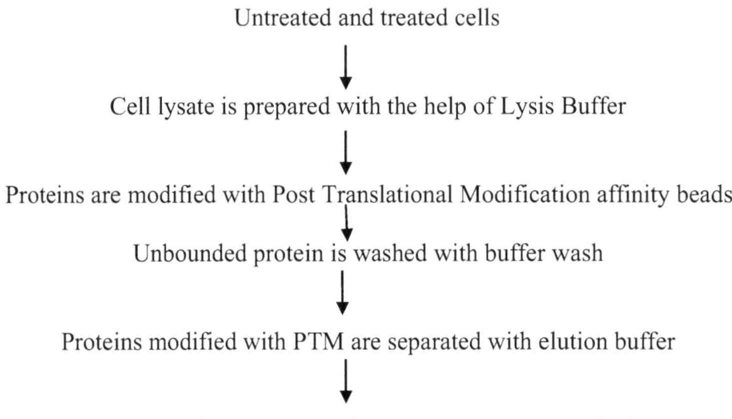

Table 5. Major Proteomics tools used to study PTMs in the proteome.

Proteomic Tools	Advantages	Disadvantages
2DE	Simple and appropriate for MS analysis	Low throughput and involves large samples
SILAC	Quantity is better for results	Reagents are quite expensive
2-DIGE	Multiplexing	Difficulty in separating low molecular weight Poor recovery of hydrophobic proteins
ICAT	Compatible with the amount of protein; quantity is invariable	PTM information is lost often

ICPL	Able to detect post-translational modification	Isotopic effect interferes retention time
Label free	Less amount of sample	Not suitable for low abundant proteins

Technical Challenges in the PTM Which Needs Immediate Attention

Bacteria undergo cellular changes to survive in changing physiological conditions by utilizing the robust weapon of PTMs. These are central mechanisms, regulating the nutritional requirements of bacteria by controlling the metabolic cascades [180, 181]. However, it is a widely accepted fact that PTMs are key features in increasing the virulence of pathogenic bacteria. Regarding this, it has become indispensable to study these PTMs to answer a basic question like what is the role of each different type of PTM and how they regulate the changes to the fluctuating environmental fluxes [182]. Hence here comes the role of mechanistic understandings of the diverse nature of PTMs. To address these questions, on one end great efforts are made by researchers to have a confluence of various technologies of genomic, proteomic, and metabolic analysis together with bioinformatics, which has led to making datasets that can track the changes at an exceptional minute details [183 - 185]. On the other end, it is full of technical challenges and difficulties which need immediate consideration to decipher the complexity of PTMs.

PTMs regulate the functioning of metabolic enzymes, but how they influence the different behaviors to adapt to fit in the cells is unclear. It is well documented how PTMs direct the enzymatic activity, but despite having the latest omics tool that has identified tons of PTMs on the bacterial proteome, machinery personalized for cellular adaptation is unanswered [31]. The challenge of using computational technology is that it provides limited information, in front of the huge experimental data availability [185]. Necessity has come to unravel the multilayer challenge posing in PTMs study, like modifications on which residues cause a change in the activity of protein behavior and influencing the physiology of the cell. Also, to have a throughput understanding we should decipher the relation between different chemical modifications participating as some of them are unauthentic, spurious and stoichiometry values differ according to metabolism [186].

Since the decades, researchers have implicated that PTMs are widespread with significant roles in the cell cycle, signaling pathways, enzymatic functioning, *etc.* Hitherto, we should not forget that these physiological processes are unique to bacteria, and hence protein modifications vary greatly with species diversity. Regarding this, a hurdle emphasized, is we cannot give a generalized function to every modification, which may differ from species to species. Firmicutes give degradative signaling if their proteins are arginine phosphorylated, which is unique to this phylum. Many bacterial species have Hanks-type kinases that are not detected in others. So to break this complexity, it is essential to know that various modifications occur in spatial or temporal domains targeting only a subpopulation. This fact can relate to specific conditions of bacteria growing on glucose and give lysine acetylation at low levels. Some kinases target serine/threonine phosphorylation highly specific in certain physiological conditions, as seen in *Bacillus subtilis* for spore formation [37]. The following Table **6** shows the currently available tools and techniques used in the PTM study, which is limited, homogenous, simplified, and needs to be considered on a priority basis and provide the composite knowledge to understand the heterogeneous bacterial populations.

Table 6. Currently available tools and their applications.

Approaches/Techniques Employed	Applications	References
Centrifugation with detergent solubilisation, Immuno Magnetic Separation (IMS) using anti-IgG-coated beads, Fluorescence-Activated Cell Sorting	Separate bacterial cells from host cells	[187 - 189]
Non-proteomics approaches: Fluorescence *In-Situ* Hybridization (FISH), Real-Time PCR, Flow Cytometry and Terminal Restriction Fragment Length Polymorphism (T-RFLP)	To study the population dynamics of binary cultures	[3, 140, 190-192]
Label free approaches or label based relative Quantitation methods iTRAQ(isobaric tagging for relative and absolute quantitation) or SILAC(stable isotope labelling with amino-acids in cell culture)	Quantification of proteome dynamics post-infection can be done	[193, 194]
2-DE followed by LC-ESI-MS	Metaproteome analysis of mixed culture	[195]
High- resolution mass spectrometry,	Bacterial infection models or bacterial co- cultures.	
Global sub cellular protein profiling	To study Subcellular distributions of proteins, and to find novel targets accountable for virulence factors	[196]
Denovo sequencing	peptide/protein identification in the absence of DNA sequences or availability of partial sequences	[197]
Pulse- chase experiments using ^{35}S-labeled methionine or cysteine	Monitor changes in protein turn over during synthesis	[140]
Multiple reaction monitoring mass spectrometry (MRM-MS)	To study the targeted proteomics like virulence factors present in Gram-positive bacterium Streptococcus pyogenes	[198]

(Table 6) cont.....

Approaches/Techniques Employed	Applications	References
Selected reaction monitoring mass spectrometry(SRM-MS)	Translate the S. pyogenes proteome	[199]
Mass-spectrometry based imaging techniques	Studying diversity in microbial society	[200]

The Lessons Learned from the Past and Giving Way to Future Solutions

Over the past few decades, technological advancement has been markedly visible in the field of PTMs proteomics, but it has also spoken of the challenges and hurdles brought forth which need to be explored. To tackle the multilayer challenges of PTMs, we need to construct interdisciplinary frameworks constituting different domains of system biology, biochemistry, and synthetic biology. This synergism will lead to creating workflows to analyze and interpret the unprecedented biological data. These domains will bring together systemic levels, multi-omics, and molecular dynamics (MD) recreation help in the PTMs modeling at the atomic levels [182].

Another difficulty of PTMs is to study their complicated communications. This can be addressed by using workflows that reconcile multiplex automated genome editing (MAGE) and metabolic modeling in metabolic engineering. These methods will help in identifying the roles of exact PTMs and will verify how PTMs adjust enzyme pathways, changing the cell phenotypes. Also, this strategy will give a clear picture of the cross talking and interplay of environment and the dynamic response of PTMS altering in the metabolomic proteome [182].

The complexity faced in assigning the generalized function of PTMs as discussed in challenges can be solved by addressing the bacterial physiological surroundings. However, this intricacy is mostly inaccessible with the current proteomic workflows, but recent improvements in mass spectrometry and sophisticated way of sample preparations have provided better insights to solve this problem. To study the regulatory roles of PTMs, we need to have an absolute level of understandings of enzymes with their target rather than a relative scale of learning. To decode the interrelated networks in the future, we need to follow a high degree of interactomic approaches to decipher the complex behavior of PTMs.

CONSENT FOR PUBLICATION

Not applicable.

CONFLICT OF INTEREST

The authors declare no conflict of interest, financial or otherwise.

ACKNOWLEDGEMENTS

Declared none.

REFERENCES

[1]   Putignani L, Del Chierico F, Petrucca A, Vernocchi P, Dallapiccola B. The human gut microbiota: a dynamic interplay with the host from birth to senescence settled during childhood. Pediatr Res 2014; 76(1): 2-10.
[http://dx.doi.org/10.1038/pr.2014.49] [PMID: 24732106]

[2]   Chowdhury F, Begum YA, Alam MM, *et al.* Concomitant enterotoxigenic *Escherichia coli* infection induces increased immune responses to *Vibrio cholerae* O1 antigens in patients with cholera in Bangladesh. Infect Immun 2010; 78(5): 2117-24.
[http://dx.doi.org/10.1128/IAI.01426-09] [PMID: 20176796]

[3]   Kluge S, Hoffmann M, Benndorf D, Rapp E, Reichl U. Proteomic tracking and analysis of a bacterial mixed culture. Proteomics 2012; 12(12): 1893-901.
[http://dx.doi.org/10.1002/pmic.201100362] [PMID: 22623171]

[4]   Yang Y, Hu M, Yu K, Zeng X, Liu X. Mass spectrometry-based proteomic approaches to study pathogenic bacteria-host interactions. Protein Cell 2015; 6(4): 265-74.
[http://dx.doi.org/10.1007/s13238-015-0136-6] [PMID: 25722051]

[5]   Tseng T-T, Tyler BM, Setubal JC. Protein secretion systems in bacterial-host associations, and their description in the Gene Ontology. BMC Microbiol 2009; 9(S1) (Suppl. 1): S2.
[http://dx.doi.org/10.1186/1471-2180-9-S1-S2] [PMID: 19278550]

[6]   Ravikumar V, Jers C, Mijakovic I. Elucidating host–pathogen interactions based on post-translational modifications using proteomics approaches. Front Microbiol 2015; 6: 1313.
[http://dx.doi.org/10.3389/fmicb.2015.01312] [PMID: 26635773]

[7]   Singh KD, Halbedel S, Görke B, Stülke J. Control of the phosphorylation state of the HPr protein of the phosphotransferase system in *Bacillus subtilis*: implication of the protein phosphatase PrpC. J Mol Microbiol Biotechnol 2007; 13(1-3): 165-71.
[http://dx.doi.org/10.1159/000103608] [PMID: 17693724]

[8]   Deribe YL, Pawson T, Dikic I. Post-translational modifications in signal integration. Nat Struct Mol Biol 2010; 17(6): 666-72.
[http://dx.doi.org/10.1038/nsmb.1842] [PMID: 20495563]

[9]   Li H, Xing X, Ding G, *et al.* SysPTM: a systematic resource for proteomic research on post-translational modifications. Mol Cell Proteomics 2009; 8(8): 1839-49.
[http://dx.doi.org/10.1074/mcp.M900030-MCP200] [PMID: 19366988]

[10]   Wang Q, Zhang Y, Yang C, *et al.* Acetylation of metabolic enzymes coordinates carbon source utilization and metabolic flux. Science 2010; 327(5968): 1004-7.
[http://dx.doi.org/10.1126/science.1179687] [PMID: 20167787]

[11]   Zhao S, Xu W, Jiang W, *et al.* Regulation of cellular metabolism by protein lysine acetylation. Science 2010; 327(5968): 1000-4.
[http://dx.doi.org/10.1126/science.1179689] [PMID: 20167786]

[12]   Grangeasse C, Stülke J, Mijakovic I. Regulatory potential of post-translational modifications in bacteria. Front Microbiol 2015; 6: 500.
[http://dx.doi.org/10.3389/fmicb.2015.00500] [PMID: 26074895]

[13]   Soufi B, Krug K, Harst A, Macek B. Characterization of the *E. coli* proteome and its modifications during growth and ethanol stress. Front Microbiol 2015; 6: 103.
[http://dx.doi.org/10.3389/fmicb.2015.00103] [PMID: 25741329]

[14]   Kunz AN, Brook I. Emerging resistant Gram-negative aerobic bacilli in hospital-acquired infections. Chemotherapy 2010; 56(6): 492-500.
[http://dx.doi.org/10.1159/000321018] [PMID: 21099222]

[15]   Rice LB. Federal funding for the study of antimicrobial resistance in nosocomial pathogens: no ESKAPE. The University of Chicago Press 2008.
[http://dx.doi.org/10.1086/533452]

[16]   Rice LB. Progress and challenges in implementing the research on ESKAPE pathogens. Infect Control Hosp Epidemiol 2010; 31(S1) (Suppl. 1): S7-S10.
[http://dx.doi.org/10.1086/655995] [PMID: 20929376]

[17]   Slama TG. Gram-negative antibiotic resistance: there is a price to pay. Crit Care 2008; 12(S4) (Suppl. 4): S4.
[http://dx.doi.org/10.1186/cc6820] [PMID: 18495061]

[18]   Dominey-Howes D, Bajorek B, Michael CA, Betteridge B, Iredell J, Labbate M. Applying the emergency risk management process to tackle the crisis of antibiotic resistance. Front Microbiol 2015; 6: 927.
[http://dx.doi.org/10.3389/fmicb.2015.00927] [PMID: 26388864]

[19]   Gupta M, Tomar RS, Kaushik S, Mishra RK, Sharma D. Effective antimicrobial activity of green ZnO nano particles of Catharanthus roseus. Front Microbiol 2018; 9: 2030.
[http://dx.doi.org/10.3389/fmicb.2018.02030] [PMID: 30233518]

[20]   Tiwari M, Roy R, Tiwari V. Screening of herbal-based bioactive extract against carbapenem-resistant strain of Acinetobacter baumannii. Microb Drug Resist 2016; 22(5): 364-71.
[http://dx.doi.org/10.1089/mdr.2015.0270] [PMID: 26910023]

[21]   Sharma D. Fluopsin C: a potential candidate against the deadly drug-resistant microbial infections in humans. Future Microbiol 2020; 15(6): 381-4.
[http://dx.doi.org/10.2217/fmb-2019-0307] [PMID: 32242758]

[22]   Sharma D, Garg A, Kumar M, Rashid F, Khan AU. Down-regulation of flagellar, fimbriae, and pili proteins in carbapenem-resistant *Klebsiella pneumoniae* (NDM-4) clinical isolates: a novel linkage to drug resistance. Front Microbiol 2019; 10: 2865.
[http://dx.doi.org/10.3389/fmicb.2019.02865] [PMID: 31921045]

[23]   Heras B, Scanlon MJ, Martin JL. Targeting virulence not viability in the search for future antibacterials. Br J Clin Pharmacol 2015; 79(2): 208-15.
[http://dx.doi.org/10.1111/bcp.12356] [PMID: 24552512]

[24]   Felise HB, Nguyen HV, Pfuetzner RA, *et al.* An inhibitor of gram-negative bacterial virulence protein secretion. Cell Host Microbe 2008; 4(4): 325-36.
[http://dx.doi.org/10.1016/j.chom.2008.08.001] [PMID: 18854237]

[25]   Hirakawa H, Tomita H. Interference of bacterial cell-to-cell communication: a new concept of antimicrobial chemotherapy breaks antibiotic resistance. Front Microbiol 2013; 4: 114.
[http://dx.doi.org/10.3389/fmicb.2013.00114] [PMID: 23720655]

[26]   Hung DT, Shakhnovich EA, Pierson E, Mekalanos JJ. Small-molecule inhibitor of *Vibrio cholerae* virulence and intestinal colonization. Science 2005; 310(5748): 670-4.
[http://dx.doi.org/10.1126/science.1116739] [PMID: 16223984]

[27]   Macek B, Forchhammer K, Hardouin J, Weber-Ban E, Grangeasse C, Mijakovic I. Protein post-translational modifications in bacteria. Nat Rev Microbiol 2019; 17(11): 651-64.
[http://dx.doi.org/10.1038/s41579-019-0243-0] [PMID: 31485032]

[28]   Latousakis D, Juge N. How sweet are our gut beneficial bacteria? A focus on protein glycosylation in Lactobacillus. Int J Mol Sci 2018; 19(1): 136.
[http://dx.doi.org/10.3390/ijms19010136] [PMID: 29301365]

[29]   Ree R, Varland S, Arnesen T. Spotlight on protein N-terminal acetylation. Exp Mol Med 2018; 50(7): 1-13.
[http://dx.doi.org/10.1038/s12276-018-0116-z] [PMID: 30054468]

[30]   Uy R, Wold F. Posttranslational covalent modification of proteins. Science 1977; 198(4320): 890-6.
[http://dx.doi.org/10.1126/science.337487] [PMID: 337487]

[31]   Cain JA, Solis N, Cordwell SJ. Beyond gene expression: the impact of protein post-translational modifications in bacteria. J Proteomics 2014; 97: 265-86.
[http://dx.doi.org/10.1016/j.jprot.2013.08.012] [PMID: 23994099]

[32]   Baenziger JU. A major step on the road to understanding a unique posttranslational modification and its role in a genetic disease. Cell 2003; 113(4): 421-2.
[http://dx.doi.org/10.1016/S0092-8674(03)00354-4] [PMID: 12757700]

[33]   Seo J, Lee K-J. Post-translational modifications and their biological functions: proteomic analysis and systematic approaches. J Biochem Mol Biol 2004; 37(1): 35-44.
[PMID: 14761301]

[34]   Olsen JV, Mann M. Status of large-scale analysis of post-translational modifications by mass spectrometry. Mol Cell Proteomics 2013; 12(12): 3444-52.
[http://dx.doi.org/10.1074/mcp.O113.034181] [PMID: 24187339]

[35]   Imber M, Pietrzyk-Brzezinska AJ, Antelmann H. Redox regulation by reversible protein S-thiolation in Gram-positive bacteria. Redox Biol 2019; 20: 130-45.
[http://dx.doi.org/10.1016/j.redox.2018.08.017] [PMID: 30308476]

[36]   Loi VV, Rossius M, Antelmann H. Redox regulation by reversible protein S-thiolation in bacteria. Front Microbiol 2015; 6: 187.
[http://dx.doi.org/10.3389/fmicb.2015.00187] [PMID: 25852656]

[37]   Macek B, Mijakovic I. Site-specific analysis of bacterial phosphoproteomes. Proteomics 2011; 11(15): 3002-11.
[http://dx.doi.org/10.1002/pmic.201100012] [PMID: 21726046]

[38]   Mann M, Jensen ON. Proteomic analysis of post-translational modifications. Nat Biotechnol 2003; 21(3): 255-61.
[http://dx.doi.org/10.1038/nbt0303-255] [PMID: 12610572]

[39]   Potel CM, Lin M-H, Heck AJR, Lemeer S. Widespread bacterial protein histidine phosphorylation revealed by mass spectrometry-based proteomics. Nat Methods 2018; 15(3): 187-90.
[http://dx.doi.org/10.1038/nmeth.4580] [PMID: 29377012]

[40]   Elsholz AK, Turgay K, Michalik S, *et al.* Global impact of protein arginine phosphorylation on the physiology of *Bacillus subtilis*. Proc Natl Acad Sci USA 2012; 109(19): 7451-6.
[http://dx.doi.org/10.1073/pnas.1117483109] [PMID: 22517742]

[41]   Junker S, Maaβ S, Otto A, *et al.* Spectral library based analysis of arginine phosphorylations in *Staphylococcus aureus*. Mol Cell Proteomics 2018; 17(2): 335-48.
[http://dx.doi.org/10.1074/mcp.RA117.000378] [PMID: 29183913]

[42]   Fortuin S, Tomazella GG, Nagaraj N, *et al.* Phosphoproteomics analysis of a clinical *Mycobacterium tuberculosis* Beijing isolate: expanding the mycobacterial phosphoproteome catalog. Front Microbiol 2015; 6: 6.
[http://dx.doi.org/10.3389/fmicb.2015.00006] [PMID: 25713560]

[43]   Nakedi KC, Nel AJ, Garnett S, Blackburn JM, Soares NC. Comparative Ser/Thr/Tyr phosphoproteomics between two mycobacterial species: the fast growing *Mycobacterium smegmatis*

and the slow growing Mycobacterium bovis BCG. Front Microbiol 2015; 6: 237.
[http://dx.doi.org/10.3389/fmicb.2015.00237] [PMID: 25904896]

[44]   Michard C, Doublet P. Post-translational modifications are key players of the Legionella pneumophila infection strategy. Front Microbiol 2015; 6: 87.
[http://dx.doi.org/10.3389/fmicb.2015.00087] [PMID: 25713573]

[45]   Shi L, Pigeonneau N, Ravikumar V, *et al.* Cross-phosphorylation of bacterial serine/threonine and tyrosine protein kinases on key regulatory residues. Front Microbiol 2014; 5: 495.
[http://dx.doi.org/10.3389/fmicb.2014.00495] [PMID: 25278935]

[46]   Shi L, Pigeonneau N, Ventroux M, *et al.* Protein-tyrosine phosphorylation interaction network in *Bacillus subtilis* reveals new substrates, kinase activators and kinase cross-talk. Front Microbiol 2014; 5: 538.
[http://dx.doi.org/10.3389/fmicb.2014.00538] [PMID: 25374563]

[47]   Dohmen RJ. SUMO protein modification. Biochimica et Biophysica Acta (BBA)-. Molecular Cell Research 2004; 1695(1-3): 113-31.

[48]   Najbauer J, Orpiszewski J, Aswad DW. Molecular aging of tubulin: accumulation of isoaspartyl sites *in vitro* and *in vivo*. Biochemistry 1996; 35(16): 5183-90.
[http://dx.doi.org/10.1021/bi953063g] [PMID: 8611502]

[49]   Pawson T. Regulation and targets of receptor tyrosine kinases. Eur J Cancer 2002; 38 (Suppl. 5): S3-S10.
[http://dx.doi.org/10.1016/S0959-8049(02)80597-4] [PMID: 12528767]

[50]   Kouzarides T. Acetylation: a regulatory modification to rival phosphorylation? EMBO J 2000; 19(6): 1176-9.
[http://dx.doi.org/10.1093/emboj/19.6.1176] [PMID: 10716917]

[51]   Schwartz JH. Ubiquitination, protein turnover, and long-term synaptic plasticity. Sci STKE 2003; 2003(190): pe26--pe.
[PMID: 12855772]

[52]   Seeler J-S, Dejean A. Nuclear and unclear functions of SUMO. Nat Rev Mol Cell Biol 2003; 4(9): 690-9.
[http://dx.doi.org/10.1038/nrm1200] [PMID: 14506472]

[53]   Botting CH, Talbot P, Paytubi S, White MF. Extensive lysine methylation in hyperthermophilic crenarchaea: potential implications for protein stability and recombinant enzymes. Archaea 2010; 2010
[http://dx.doi.org/10.1155/2010/106341]

[54]   Ge W, Wolf A, Feng T, *et al.* Oxygenase-catalyzed ribosome hydroxylation occurs in prokaryotes and humans. Nat Chem Biol 2012; 8(12): 960-2.
[http://dx.doi.org/10.1038/nchembio.1093] [PMID: 23103944]

[55]   Bourret RB, Borkovich KA, Simon MI. Signal transduction pathways involving protein phosphorylation in prokaryotes. Annu Rev Biochem 1991; 60(1): 401-41.
[http://dx.doi.org/10.1146/annurev.bi.60.070191.002153] [PMID: 1883200]

[56]   Stock AM, Robinson VL, Goudreau PN. Two-component signal transduction. Annu Rev Biochem 2000; 69(1): 183-215.
[http://dx.doi.org/10.1146/annurev.biochem.69.1.183] [PMID: 10966457]

[57]   Ren J, Sang Y, Lu J, Yao Y-F. Protein acetylation and its role in bacterial virulence. Trends Microbiol 2017; 25(9): 768-79.
[http://dx.doi.org/10.1016/j.tim.2017.04.001] [PMID: 28462789]

[58]   Sang Y, Ren J, Qin R, *et al.* Acetylation regulating protein stability and DNA-binding ability of HilD, thus modulating *Salmonella Typhimurium* virulence. J Infect Dis 2017; 216(8): 1018-26.
[http://dx.doi.org/10.1093/infdis/jix102] [PMID: 28329249]

[59] Zhang Q, Zhou A, Li S, *et al.* Reversible lysine acetylation is involved in DNA replication initiation by regulating activities of initiator DnaA in *Escherichia coli.* Sci Rep 2016; 6(1): 30837.
[http://dx.doi.org/10.1038/srep30837] [PMID: 27484197]

[60] Sangith N, Kumar S, Sankaran K. Evidence to Suggest Bacterial Lipoprotein Diacylglyceryl Transferase (Lgt) is a Weakly Associated Inner Membrane Protein. J Membr Biol 2019; 252(6): 563-75.
[http://dx.doi.org/10.1007/s00232-019-00076-3] [PMID: 31256204]

[61] Varki A. Biological roles of glycans. Glycobiology 2017; 27(1): 3-49.
[http://dx.doi.org/10.1093/glycob/cww086] [PMID: 27558841]

[62] Collier RJ, Cole HA. Diphtheria toxin subunit active *in vitro.* Science 1969; 164(3884): 1179-81.
[http://dx.doi.org/10.1126/science.164.3884.1179] [PMID: 4305968]

[63] Ribet D, Cossart P. Post-translational modifications in host cells during bacterial infection. FEBS Lett 2010; 584(13): 2748-58.
[http://dx.doi.org/10.1016/j.febslet.2010.05.012] [PMID: 20493189]

[64] Ribet D, Cossart P. Pathogen-mediated posttranslational modifications: A re-emerging field. Cell 2010; 143(5): 694-702.
[http://dx.doi.org/10.1016/j.cell.2010.11.019] [PMID: 21111231]

[65] Starai VJ, Escalante-Semerena JC. Identification of the protein acetyltransferase (Pat) enzyme that acetylates acetyl-CoA synthetase in *Salmonella enterica.* J Mol Biol 2004; 340(5): 1005-12.
[http://dx.doi.org/10.1016/j.jmb.2004.05.010] [PMID: 15236963]

[66] Weinert BT, Iesmantavicius V, Wagner SA, *et al.* Acetyl-phosphate is a critical determinant of lysine acetylation in *E. coli.* Mol Cell 2013; 51(2): 265-72.
[http://dx.doi.org/10.1016/j.molcel.2013.06.003] [PMID: 23830618]

[67] Butler CA, Veith PD, Nieto MF, Dashper SG, Reynolds EC. Lysine acetylation is a common post-translational modification of key metabolic pathway enzymes of the anaerobe Porphyromonas gingivalis. J Proteomics 2015; 128: 352-64.
[http://dx.doi.org/10.1016/j.jprot.2015.08.015] [PMID: 26341301]

[68] Linton D, Allan E, Karlyshev AV, Cronshaw AD, Wren BW. Identification of N-acetylgalactosamin--containing glycoproteins PEB3 and CgpA in Campylobacter jejuni. Mol Microbiol 2002; 43(2): 497-508.
[http://dx.doi.org/10.1046/j.1365-2958.2002.02762.x] [PMID: 11985725]

[69] Vik A, Aas FE, Anonsen JH, *et al.* Broad spectrum O-linked protein glycosylation in the human pathogen *Neisseria gonorrhoeae.* Proc Natl Acad Sci USA 2009; 106(11): 4447-52.
[http://dx.doi.org/10.1073/pnas.0809504106] [PMID: 19251655]

[70] Lithgow KV, Scott NE, Iwashkiw JA, *et al.* A general protein O-glycosylation system within the Burkholderia cepacia complex is involved in motility and virulence. Mol Microbiol 2014; 92(1): 116-37.
[http://dx.doi.org/10.1111/mmi.12540] [PMID: 24673753]

[71] García-Fontana C, Corral Lugo A, Krell T. Specificity of the CheR2 methyltransferase in *Pseudomonas aeruginosa* is directed by a C-terminal pentapeptide in the McpB chemoreceptor. Sci Signal 2014; 7(320): ra34-[-ra.].
[http://dx.doi.org/10.1126/scisignal.2004849] [PMID: 24714571]

[72] Sharma G, Upadhyay S, Srilalitha M, Nandicoori VK, Khosla S. The interaction of mycobacterial protein Rv2966c with host chromatin is mediated through non-CpG methylation and histone H3/H4 binding. Nucleic Acids Res 2015; 43(8): 3922-37.
[http://dx.doi.org/10.1093/nar/gkv261] [PMID: 25824946]

[73] Abeykoon A, Wang G, Chao C-C, *et al.* Multimethylation of Rickettsia OmpB catalyzed by lysine methyltransferases. J Biol Chem 2014; 289(11): 7691-701.

[http://dx.doi.org/10.1074/jbc.M113.535567] [PMID: 24497633]

[74]   Owings JP, Kuiper EG, Prezioso SM, *et al. Pseudomonas aeruginosa* EftM is a thermoregulated methyltransferase. J Biol Chem 2016; 291(7): 3280-90.
[http://dx.doi.org/10.1074/jbc.M115.706853] [PMID: 26677219]

[75]   Yaseen I, Choudhury M, Sritharan M, Khosla S. Histone methyltransferase SUV39H1 participates in host defense by methylating mycobacterial histone-like protein HupB. EMBO J 2018; 37(2): 183-200.
[http://dx.doi.org/10.15252/embj.201796918] [PMID: 29170282]

[76]   Bodenmiller B, Wanka S, Kraft C, *et al.* Phosphoproteomic analysis reveals interconnected system-wide responses to perturbations of kinases and phosphatases in yeast. Sci Signal 2010; 3(153): rs4--rs.].
[http://dx.doi.org/10.1126/scisignal.2001182] [PMID: 21177495]

[77]   Zorina A, Stepanchenko N, Novikova GV, *et al.* Eukaryotic-like Ser/Thr protein kinases SpkC/F/K are involved in phosphorylation of GroES in the Cyanobacterium synechocystis. DNA Res 2011; 18(3): 137-51.
[http://dx.doi.org/10.1093/dnares/dsr006] [PMID: 21551175]

[78]   Gnad F, Gunawardena J, Mann M. PHOSIDA 2011: the posttranslational modification database. Nucleic acids research 2010; 39(suppl_1): D253-60.

[79]   Dinkel H, Chica C, Gould CM, Jensen LJ, Gibson TJ, *et al.* Phospho. ELM: a database of phosphorylation sites—update 2011. Nucleic Acids Research 2010; 39(suppl_1): D261-7.

[80]   Hornbeck PV, Chabra I, Kornhauser JM, Skrzypek E, Zhang B. PhosphoSite: A bioinformatics resource dedicated to physiological protein phosphorylation. Proteomics 2004; 4(6): 1551-61.
[http://dx.doi.org/10.1002/pmic.200300772] [PMID: 15174125]

[81]   Liao S, Shang Q, Zhang X, Zhang J, Xu C, Tu X. Pup, a prokaryotic ubiquitin-like protein, is an intrinsically disordered protein. Biochem J 2009; 422(2): 207-15.
[http://dx.doi.org/10.1042/BJ20090738] [PMID: 19580545]

[82]   Burns KE, Liu W-T, Boshoff HI, Dorrestein PC, Barry CE III. Proteasomal protein degradation in Mycobacteria is dependent upon a prokaryotic ubiquitin-like protein. J Biol Chem 2009; 284(5): 3069-75.
[http://dx.doi.org/10.1074/jbc.M808032200] [PMID: 19028679]

[83]   De Mot R, Nagy I, Walz J, Baumeister W. Proteasomes and other self-compartmentalizing proteases in prokaryotes. Trends Microbiol 1999; 7(2): 88-92.
[http://dx.doi.org/10.1016/S0966-842X(98)01432-2] [PMID: 10081087]

[84]   Ohman DE. Molecular genetics of exopolysaccharide production by mucoid *Pseudomonas aeruginosa.* Eur J Clin Microbiol 1986; 5(1): 6-10.
[http://dx.doi.org/10.1007/BF02013452] [PMID: 3009178]

[85]   Allesen-Holm M, Barken KB, Yang L, *et al.* A characterization of DNA release in *Pseudomonas aeruginosa* cultures and biofilms. Mol Microbiol 2006; 59(4): 1114-28.
[http://dx.doi.org/10.1111/j.1365-2958.2005.05008.x] [PMID: 16430688]

[86]   Smidsrod O, Draget K. Chemistry and physical properties of alginates. 1996.

[87]   Nivens DE, Ohman DE, Williams J, Franklin MJ. Role of alginate and its O acetylation in formation of *Pseudomonas aeruginosa* microcolonies and biofilms. J Bacteriol 2001; 183(3): 1047-57.
[http://dx.doi.org/10.1128/JB.183.3.1047-1057.2001] [PMID: 11208804]

[88]   Mai GT, Seow WK, Pier GB, McCormack JG, Thong YH. Suppression of lymphocyte and neutrophil functions by *Pseudomonas aeruginosa* mucoid exopolysaccharide (alginate): reversal by physicochemical, alginase, and specific monoclonal antibody treatments. Infect Immun 1993; 61(2): 559-64.
[http://dx.doi.org/10.1128/IAI.61.2.559-564.1993] [PMID: 8423085]

[89]   Pier GB, Coleman F, Grout M, Franklin M, Ohman DE. Role of alginate O acetylation in resistance of mucoid *Pseudomonas aeruginosa* to opsonic phagocytosis. Infect Immun 2001; 69(3): 1895-901.
[http://dx.doi.org/10.1128/IAI.69.3.1895-1901.2001] [PMID: 11179370]

[90]   Donati I, Holtan S, Mørch YA, Borgogna M, Dentini M, Skjåk-Braek G. New hypothesis on the role of alternating sequences in calcium-alginate gels. Biomacromolecules 2005; 6(2): 1031-40.
[http://dx.doi.org/10.1021/bm049306e] [PMID: 15762675]

[91]   Sarkisova S, Patrauchan MA, Berglund D, Nivens DE, Franklin MJ. Calcium-induced virulence factors associated with the extracellular matrix of mucoid *Pseudomonas aeruginosa* biofilms. J Bacteriol 2005; 187(13): 4327-37.
[http://dx.doi.org/10.1128/JB.187.13.4327-4337.2005] [PMID: 15968041]

[92]   Franklin MJ, Nivens DE, Weadge JT, Howell PL. Biosynthesis of the *Pseudomonas aeruginosa* Extracellular Polysaccharides, Alginate, Pel, and Psl. Front Microbiol 2011; 2: 167.
[http://dx.doi.org/10.3389/fmicb.2011.00167] [PMID: 21991261]

[93]   Colvin KM, Alnabelseya N, Baker P, Whitney JC, Howell PL, Parsek MR. PelA deacetylase activity is required for Pel polysaccharide synthesis in *Pseudomonas aeruginosa*. J Bacteriol 2013; 195(10): 2329-39.
[http://dx.doi.org/10.1128/JB.02150-12] [PMID: 23504011]

[94]   Cescutti P, Foschiatti M, Furlanis L, Lagatolla C, Rizzo R. Isolation and characterisation of the biological repeating unit of cepacian, the exopolysaccharide produced by bacteria of the Burkholderia cepacia complex. Carbohydr Res 2010; 345(10): 1455-60.
[http://dx.doi.org/10.1016/j.carres.2010.03.029] [PMID: 20409536]

[95]   Ferreira AS, Silva IN, Oliveira VH, Cunha R, Moreira LM. Insights into the role of extracellular polysaccharides in Burkholderia adaptation to different environments. Front Cell Infect Microbiol 2011; 1: 16.
[http://dx.doi.org/10.3389/fcimb.2011.00016] [PMID: 22919582]

[96]   Galyov EE, Håkansson S, Forsberg A, Wolf-Watz H. A secreted protein kinase of Yersinia pseudotuberculosis is an indispensable virulence determinant. Nature 1993; 361(6414): 730-2.
[http://dx.doi.org/10.1038/361730a0] [PMID: 8441468]

[97]   Prabakaran S, Lippens G, Steen H, Gunawardena J. Post-translational modification: nature's escape from genetic imprisonment and the basis for dynamic information encoding. Wiley Interdiscip Rev Syst Biol Med 2012; 4(6): 565-83.
[http://dx.doi.org/10.1002/wsbm.1185] [PMID: 22899623]

[98]   Walsh CT, Garneau-Tsodikova S, Gatto GJ Jr. Protein posttranslational modifications: the chemistry of proteome diversifications. Angew Chem Int Ed Engl 2005; 44(45): 7342-72.
[http://dx.doi.org/10.1002/anie.200501023] [PMID: 16267872]

[99]   Manning G, Whyte DB, Martinez R, Hunter T, Sudarsanam S. The protein kinase complement of the human genome. Science 2002; 298(5600): 1912-34.
[http://dx.doi.org/10.1126/science.1075762] [PMID: 12471243]

[100] Mogensen TH. Pathogen recognition and inflammatory signaling in innate immune defenses. Clin Microbiol Rev 2009; 22(2): 240-73.
[http://dx.doi.org/10.1128/CMR.00046-08] [PMID: 19366914]

[101] Aderem A, Ulevitch RJ. Toll-like receptors in the induction of the innate immune response. Nature 2000; 406(6797): 782-7.
[http://dx.doi.org/10.1038/35021228] [PMID: 10963608]

[102] Stimson E, Virji M, Makepeace K, *et al.* Meningococcal pilin: a glycoprotein substituted with digalactosyl 2,4-diacetamido-2,4,6-trideoxyhexose. Mol Microbiol 1995; 17(6): 1201-14.
[http://dx.doi.org/10.1111/j.1365-2958.1995.mmi_17061201.x] [PMID: 8594338]

[103] Grass S, Buscher AZ, Swords WE, *et al.* The Haemophilus influenzae HMW1 adhesin is glycosylated

in a process that requires HMW1C and phosphoglucomutase, an enzyme involved in lipooligosaccharide biosynthesis. Mol Microbiol 2003; 48(3): 737-51.
[http://dx.doi.org/10.1046/j.1365-2958.2003.03450.x] [PMID: 12694618]

[104] Szymanski CM, Yao R, Ewing CP, Trust TJ, Guerry P. Evidence for a system of general protein glycosylation in Campylobacter jejuni. Mol Microbiol 1999; 32(5): 1022-30.
[http://dx.doi.org/10.1046/j.1365-2958.1999.01415.x] [PMID: 10361304]

[105] Dobos KM, Khoo K-H, Swiderek KM, Brennan PJ, Belisle JT. Definition of the full extent of glycosylation of the 45-kilodalton glycoprotein of *Mycobacterium tuberculosis*. J Bacteriol 1996; 178(9): 2498-506.
[http://dx.doi.org/10.1128/JB.178.9.2498-2506.1996] [PMID: 8626314]

[106] Wu H, Mintz KP, Ladha M, Fives-Taylor PM. Isolation and characterization of Fap1, a fimbriae-associated adhesin of *Streptococcus parasanguis* FW213. Mol Microbiol 1998; 28(3): 487-500.
[http://dx.doi.org/10.1046/j.1365-2958.1998.00805.x] [PMID: 9632253]

[107] Benz I, Schmidt MA. Glycosylation with heptose residues mediated by the aah gene product is essential for adherence of the AIDA-I adhesin. Mol Microbiol 2001; 40(6): 1403-13.
[http://dx.doi.org/10.1046/j.1365-2958.2001.02487.x] [PMID: 11442838]

[108] Tytgat HL, van Teijlingen NH, Sullan RMA, *et al.* Probiotic gut microbiota isolate interacts with dendritic cells *via* glycosylated heterotrimeric pili. PLoS One 2016; 11(3): e0151824.
[http://dx.doi.org/10.1371/journal.pone.0151824] [PMID: 26985831]

[109] Varki A. Biological roles of oligosaccharides: all of the theories are correct. Glycobiology 1993; 3(2): 97-130.
[http://dx.doi.org/10.1093/glycob/3.2.97] [PMID: 8490246]

[110] Tytgat HLP, de Vos WM. Sugar coating the envelope: glycoconjugates for microbe–host crosstalk. Trends Microbiol 2016; 24(11): 853-61.
[http://dx.doi.org/10.1016/j.tim.2016.06.004] [PMID: 27374775]

[111] Jank T, Aktories K. Structure and mode of action of clostridial glucosylating toxins: the ABCD model. Trends Microbiol 2008; 16(5): 222-9.
[http://dx.doi.org/10.1016/j.tim.2008.01.011] [PMID: 18394902]

[112] Carabetta VJ, Cristea IM. Regulation, function, and detection of protein acetylation in bacteria. J Bacteriol 2017; 199(16): e00107-17.
[http://dx.doi.org/10.1128/JB.00107-17] [PMID: 28439035]

[113] Striebel F, Hunkeler M, Summer H, Weber-Ban E. The mycobacterial Mpa-proteasome unfolds and degrades pupylated substrates by engaging Pup's N-terminus. EMBO J 2010; 29(7): 1262-71.
[http://dx.doi.org/10.1038/emboj.2010.23] [PMID: 20203624]

[114] Bray BA, Sutcliffe IC, Harrington DJ. Impact of lgt mutation on lipoprotein biosynthesis and *in vitro* phenotypes of *Streptococcus agalactiae*. Microbiology (Reading) 2009; 155(Pt 5): 1451-8.
[http://dx.doi.org/10.1099/mic.0.025213-0] [PMID: 19383708]

[115] Yang S-J, Rice KC, Brown RJ, *et al.* A LysR-type regulator, CidR, is required for induction of the *Staphylococcus aureus* cidABC operon. J Bacteriol 2005; 187(17): 5893-900.
[http://dx.doi.org/10.1128/JB.187.17.5893-5900.2005] [PMID: 16109930]

[116] Kucukyildirim S, Long H, Sung W, Miller SF, Doak TG, Lynch M. The rate and spectrum of spontaneous mutations in *Mycobacterium smegmatis*, a bacterium naturally devoid of the postreplicative mismatch repair pathway. G3: Genes, Genomes. G3 (Bethesda) 2016; 6(7): 2157-63.
[http://dx.doi.org/10.1534/g3.116.030130] [PMID: 27194804]

[117] Mujtaba S, Winer BY, Jaganathan A, *et al.* Anthrax SET protein: a potential virulence determinant that epigenetically represses NF-κB activation in infected macrophages. J Biol Chem 2013; 288(32): 23458-72.
[http://dx.doi.org/10.1074/jbc.M113.467696] [PMID: 23720780]

[118]   Dilweg IW, Dame RT. Post-translational modification of nucleoid-associated proteins: an extra layer of functional modulation in bacteria? Biochem Soc Trans 2018; 46(5): 1381-92.
[http://dx.doi.org/10.1042/BST20180488] [PMID: 30287510]

[119]   Zhang W, Sun J, Cao H, *et al.* Post-translational modifications are enriched within protein functional groups important to bacterial adaptation within a deep-sea hydrothermal vent environment. Microbiome 2016; 4(1): 49.
[http://dx.doi.org/10.1186/s40168-016-0194-x] [PMID: 27600525]

[120]   Małecki J, Dahl H-A, Moen A, Davydova E, Falnes PØ. The METTL20 homologue from *Agrobacterium tumefaciens* is a dual specificity protein-lysine methyltransferase that targets ribosomal protein L7/L12 and the β subunit of electron transfer flavoprotein (ETFβ). J Biol Chem 2016; 291(18): 9581-95.
[http://dx.doi.org/10.1074/jbc.M115.709261] [PMID: 26929405]

[121]   Xia Y, Niu Y, Cui J, *et al.* The Helicase Activity of Hyperthermophilic Archaeal MCM is Enhanced at High Temperatures by Lysine Methylation. Front Microbiol 2015; 6: 1247.
[http://dx.doi.org/10.3389/fmicb.2015.01247] [PMID: 26617586]

[122]   Cao X-J, Dai J, Xu H, *et al.* High-coverage proteome analysis reveals the first insight of protein modification systems in the pathogenic spirochete Leptospira interrogans. Cell Res 2010; 20(2): 197-210.
[http://dx.doi.org/10.1038/cr.2009.127] [PMID: 19918266]

[123]   Abeykoon AH, Chao C-C, Wang G, Gucek M, Yang DC, Ching W-M. Two protein lysine methyltransferases methylate outer membrane protein B from Rickettsia. J Bacteriol 2012; 194(23): 6410-8.
[http://dx.doi.org/10.1128/JB.01379-12] [PMID: 23002218]

[124]   Barbier M, Owings JP, Martínez-Ramos I, *et al.* Lysine trimethylation of EF-Tu mimics platelet-activating factor to initiate *Pseudomonas aeruginosa* pneumonia. MBio 2013; 4(3): e00207-13.
[http://dx.doi.org/10.1128/mBio.00207-13] [PMID: 23653444]

[125]   Brown MT, Delalez NJ, Armitage JP. Protein dynamics and mechanisms controlling the rotational behaviour of the bacterial flagellar motor. Curr Opin Microbiol 2011; 14(6): 734-40.
[http://dx.doi.org/10.1016/j.mib.2011.09.009] [PMID: 21955888]

[126]   Wadhams GH, Armitage JP. Making sense of it all: bacterial chemotaxis. Nat Rev Mol Cell Biol 2004; 5(12): 1024-37.
[http://dx.doi.org/10.1038/nrm1524] [PMID: 15573139]

[127]   Lybarger SR, Maddock JR. Clustering of the chemoreceptor complex in *Escherichia coli* is independent of the methyltransferase CheR and the methylesterase CheB. J Bacteriol 1999; 181(17): 5527-9.
[http://dx.doi.org/10.1128/JB.181.17.5527-5529.1999] [PMID: 10464232]

[128]   Eshghi A, Pinne M, Haake DA, Zuerner RL, Frank A, Cameron CE. Methylation and *in vivo* expression of the surface-exposed Leptospira interrogans outer-membrane protein OmpL32. Microbiology (Reading) 2012; 158(Pt 3): 622-35.
[http://dx.doi.org/10.1099/mic.0.054767-0] [PMID: 22174381]

[129]   Kuhn ML, Zemaitaitis B, Hu LI, *et al.* Structural, kinetic and proteomic characterization of acetyl phosphate-dependent bacterial protein acetylation. PLoS One 2014; 9(4): e94816.
[http://dx.doi.org/10.1371/journal.pone.0094816] [PMID: 24756028]

[130]   Ghosh S, Padmanabhan B, Anand C, Nagaraja V. Lysine acetylation of the *Mycobacterium tuberculosis* HU protein modulates its DNA binding and genome organization. Mol Microbiol 2016; 100(4): 577-88.
[http://dx.doi.org/10.1111/mmi.13339] [PMID: 26817737]

[131]   Tu S, Guo S-J, Chen C-S, *et al.* YcgC represents a new protein deacetylase family in prokaryotes.

eLife 2015; 4: e05322.
[http://dx.doi.org/10.7554/eLife.05322] [PMID: 26716769]

[132] Hentchel KL, Escalante-Semerena JC. Acylation of biomolecules in prokaryotes: a widespread strategy for the control of biological function and metabolic stress. Microbiol Mol Biol Rev 2015; 79(3): 321-46.
[http://dx.doi.org/10.1128/MMBR.00020-15] [PMID: 26179745]

[133] Castaño-Cerezo S, Bernal V, Blanco-Catalá J, Iborra JL, Cánovas M. cAMP-CRP co-ordinates the expression of the protein acetylation pathway with central metabolism in *Escherichia coli*. Mol Microbiol 2011; 82(5): 1110-28.
[http://dx.doi.org/10.1111/j.1365-2958.2011.07873.x] [PMID: 22059728]

[134] Liimatta K, Flaherty E, Ro G, Nguyen DK, Prado C, Purdy AE. A putative acetylation system in *Vibrio cholerae* modulates virulence in arthropod hosts. Appl Environ Microbiol 2018; 84(21): e01113-8.
[http://dx.doi.org/10.1128/AEM.01113-18] [PMID: 30143508]

[135] Liu W, Tan Y, Cao S, *et al.* Protein acetylation mediated by YfiQ and CobB is involved in the virulence and stress response of Yersinia pestis. Infect Immun 2018; 86(6): e00224-18.
[http://dx.doi.org/10.1128/IAI.00224-18] [PMID: 29610260]

[136] Ren J, Sang Y, Ni J, *et al.* Acetylation regulates survival of *Salmonella enterica* serovar Typhimurium under acid stress. Appl Environ Microbiol 2015; 81(17): 5675-82.
[http://dx.doi.org/10.1128/AEM.01009-15] [PMID: 26070677]

[137] Ma Q, Wood TK. Protein acetylation in prokaryotes increases stress resistance. Biochem Biophys Res Commun 2011; 410(4): 846-51.
[http://dx.doi.org/10.1016/j.bbrc.2011.06.076] [PMID: 21703240]

[138] AbouElfetouh A, Kuhn ML, Hu LI, *et al.* The *E. coli* sirtuin CobB shows no preference for enzymatic and nonenzymatic lysine acetylation substrate sites. MicrobiologyOpen 2015; 4(1): 66-83.
[http://dx.doi.org/10.1002/mbo3.223] [PMID: 25417765]

[139] Nakayasu ES, Burnet MC, Walukiewicz HE, *et al.* Ancient regulatory role of lysine acetylation in central metabolism. MBio 2017; 8(6): e01894-17.
[http://dx.doi.org/10.1128/mBio.01894-17] [PMID: 29184018]

[140] Schmidt F, Völker U. Proteome analysis of host-pathogen interactions: Investigation of pathogen responses to the host cell environment. Proteomics 2011; 11(15): 3203-11.
[http://dx.doi.org/10.1002/pmic.201100158] [PMID: 21710565]

[141] Trentini DB, Suskiewicz MJ, Heuck A, *et al.* Arginine phosphorylation marks proteins for degradation by a Clp protease. Nature 2016; 539(7627): 48-53.
[http://dx.doi.org/10.1038/nature20122] [PMID: 27749819]

[142] Zhou B, Semanjski M, Orlovetskie N, *et al.* Arginine dephosphorylation propels spore germination in bacteria. Proc Natl Acad Sci USA 2019; 116(28): 14228-37.
[http://dx.doi.org/10.1073/pnas.1817742116] [PMID: 31221751]

[143] Yuan J, Jin F, Glatter T, Sourjik V. Osmosensing by the bacterial PhoQ/PhoP two-component system. Proc Natl Acad Sci USA 2017; 114(50): E10792-8.
[http://dx.doi.org/10.1073/pnas.1717272114] [PMID: 29183977]

[144] Deng L, Mu R, Weston TA, Spencer BL, Liles RP, Doran KS. Characterization of a two-component system transcriptional regulator, LtdR, that impacts group B streptococcal colonization and disease. Infect Immun 2018; 86(7): e00822-17.
[http://dx.doi.org/10.1128/IAI.00822-17] [PMID: 29685987]

[145] Namugenyi SB, Aagesen AM, Elliott SR, Tischler AD. *Mycobacterium tuberculosis* PhoY proteins promote persister formation by mediating Pst/SenX3-RegX3 phosphate sensing. MBio 2017; 8(4): e00494-17.

[http://dx.doi.org/10.1128/mBio.00494-17] [PMID: 28698272]

[146]   Yadav GS, Ravala SK, Malhotra N, Chakraborti PK. Phosphorylation modulates catalytic activity of mycobacterial sirtuins. Front Microbiol 2016; 7: 677.
[http://dx.doi.org/10.3389/fmicb.2016.00677] [PMID: 27242704]

[147]   Lee WL, Singaravelu P, Wee S, *et al.* Mechanisms of Yersinia YopO kinase substrate specificity. Sci Rep 2017; 7(1): 39998.
[http://dx.doi.org/10.1038/srep39998] [PMID: 28051168]

[148]   van Staalduinen LM, Jia Z. Post-translational hydroxylation by 2OG/Fe(II)-dependent oxygenases as a novel regulatory mechanism in bacteria. Front Microbiol 2015; 5: 798.
[http://dx.doi.org/10.3389/fmicb.2014.00798] [PMID: 25642226]

[149]   Sobocińska J, Roszczenko-Jasińska P, Ciesielska A, Kwiatkowska K. Protein palmitoylation and its role in bacterial and viral infections. Front Immunol 2018; 8: 2003.
[http://dx.doi.org/10.3389/fimmu.2017.02003] [PMID: 29403483]

[150]   Spera JM, Guaimas F, Corvi MM, Ugalde JE. Brucella hijacks host-mediated palmitoylation to stabilize and localize PrpA to the plasma membrane. Infect Immun 2018; 86(11): e00402-18.
[http://dx.doi.org/10.1128/IAI.00402-18] [PMID: 30126897]

[151]   da Silva RAG, Churchward CP, Karlyshev AV, *et al.* The role of apolipoprotein N-acyl transferase, Lnt, in the lipidation of factor H binding protein of Neisseria meningitidis strain MC58 and its potential as a drug target. Br J Pharmacol 2017; 174(14): 2247-60.
[http://dx.doi.org/10.1111/bph.13660] [PMID: 27784136]

[152]   Nguyen JQ, Gilley RP, Zogaj X, Rodriguez SA, Klose KE. Lipidation of the FPI protein IglE contributes to Francisella tularensis ssp. novicida intramacrophage replication and virulence. Pathog Dis 2014; 72(1): 10-8.
[http://dx.doi.org/10.1111/2049-632X.12167] [PMID: 24616435]

[153]   Wenzel M, Schriek P, Prochnow P, Albada HB, Metzler-Nolte N, Bandow JE. Influence of lipidation on the mode of action of a small RW-rich antimicrobial peptide. Biochim Biophys Acta 2016; 1858(5): 1004-11.
[http://dx.doi.org/10.1016/j.bbamem.2015.11.009] [PMID: 26603779]

[154]   Veyron S, Oliva G, Rolando M, Buchrieser C, Peyroche G, Cherfils JA. Ca^{2+}-regulated deAMPylation switch in human and bacterial FIC proteins. bioRxiv 2018; 323253.

[155]   DiNovo AA, Schey KL, Vachon WS, McGuffie EM, Olson JC, Vincent TS. ADP-ribosylation of cyclophilin A by *Pseudomonas aeruginosa* exoenzyme S. Biochemistry 2006; 45(14): 4664-73.
[http://dx.doi.org/10.1021/bi0513554] [PMID: 16584201]

[156]   Cain JA, Dale AL, Niewold P, *et al.* Proteomics reveals multiple phenotypes associated with N-linked glycosylation in Campylobacter jejuni. Mol Cell Proteomics 2019; 18(4): 715-34.
[http://dx.doi.org/10.1074/mcp.RA118.001199] [PMID: 30617158]

[157]   Lassak J, Keilhauer EC, Fürst M, *et al.* Arginine-rhamnosylation as new strategy to activate translation elongation factor P. Nat Chem Biol 2015; 11(4): 266-70.
[http://dx.doi.org/10.1038/nchembio.1751] [PMID: 25686373]

[158]   Eichler J, Koomey M. Sweet new roles for protein glycosylation in prokaryotes. Trends Microbiol 2017; 25(8): 662-72.
[http://dx.doi.org/10.1016/j.tim.2017.03.001] [PMID: 28341406]

[159]   Cha J, Mobashery S. Lysine N(ζ)-decarboxylation in the BlaR1 protein from *Staphylococcus aureus* at the root of its function as an antibiotic sensor. J Am Chem Soc 2007; 129(13): 3834-5.
[http://dx.doi.org/10.1021/ja070472e] [PMID: 17343387]

[160]   Meulenbroek EM, Paspaleva K, Thomassen EA, Abrahams JP, Goosen N, Pannu NS. Involvement of a carboxylated lysine in UV damage endonuclease. Protein Sci 2009; 18(3): 549-58.
[http://dx.doi.org/10.1002/pro.54] [PMID: 19241382]

[161] Seth D, Hausladen A, Wang Y-J, Stamler JS. Endogenous protein S-Nitrosylation in *E. coli*: regulation by OxyR. Science 2012; 336(6080): 470-3.
[http://dx.doi.org/10.1126/science.1215643] [PMID: 22539721]

[162] Compton CL, Fernandopulle MS, Nagari RT, Sello JK. Genetic and proteomic analyses of pupylation in Streptomyces coelicolor. J Bacteriol 2015; 197(17): 2747-53.
[http://dx.doi.org/10.1128/JB.00302-15] [PMID: 26031910]

[163] Fimlaid KA, Jensen O, Donnelly ML, Francis MB, Sorg JA, Shen A. Identification of a novel lipoprotein regulator of Clostridium difficile spore germination. PLoS Pathog 2015; 11(10): e1005239.
[http://dx.doi.org/10.1371/journal.ppat.1005239] [PMID: 26496694]

[164] Elharar Y, Roth Z, Hermelin I, *et al.* Survival of mycobacteria depends on proteasome-mediated amino acid recycling under nutrient limitation. EMBO J 2014; 33(16): 1802-14.
[http://dx.doi.org/10.15252/embj.201387076] [PMID: 24986881]

[165] Samanovic MI, Tu S, Novák O, *et al.* Proteasomal control of cytokinin synthesis protects *Mycobacterium tuberculosis* against nitric oxide. Mol Cell 2015; 57(6): 984-94.
[http://dx.doi.org/10.1016/j.molcel.2015.01.024] [PMID: 25728768]

[166] Sharma D, Kumar B, Lata M, *et al.* Comparative proteomic analysis of aminoglycosides resistant and susceptible *Mycobacterium tuberculosis* clinical isolates for exploring potential drug targets. PLoS One 2015; 10(10): e0139414.
[http://dx.doi.org/10.1371/journal.pone.0139414] [PMID: 26436944]

[167] Suskiewicz MJ, Hajdusits B, Beveridge R, *et al.* Structure of McsB, a protein kinase for regulated arginine phosphorylation. Nat Chem Biol 2019; 15(5): 510-8.
[http://dx.doi.org/10.1038/s41589-019-0265-y] [PMID: 30962626]

[168] Gross R, Aricò B, Rappuoli R. Families of bacterial signal-transducing proteins. Mol Microbiol 1989; 3(11): 1661-7.
[http://dx.doi.org/10.1111/j.1365-2958.1989.tb00152.x] [PMID: 2559300]

[169] Vo CD, Shebert HL, Zikovich S, *et al.* Repurposing Hsp90 inhibitors as antibiotics targeting histidine kinases. Bioorg Med Chem Lett 2017; 27(23): 5235-44.
[http://dx.doi.org/10.1016/j.bmcl.2017.10.036] [PMID: 29110989]

[170] Bae H-J, Lee H-N, Baek M-N, *et al.* Inhibition of the DevSR two-component system by overexpression of *Mycobacterium tuberculosis* PknB in *Mycobacterium smegmatis*. Mol Cells 2017; 40(9): 632-42.
[PMID: 28843272]

[171] Canova MJ, Molle V. Bacterial serine/threonine protein kinases in host-pathogen interactions. J Biol Chem 2014; 289(14): 9473-9.
[http://dx.doi.org/10.1074/jbc.R113.529917] [PMID: 24554701]

[172] Charlton TM, Kovacs-Simon A, Michell SL, Fairweather NF, Tate EW. Quantitative lipoproteomics in Clostridium difficile reveals a role for lipoproteins in sporulation. Chem Biol 2015; 22(11): 1562-73.
[http://dx.doi.org/10.1016/j.chembiol.2015.10.006] [PMID: 26584780]

[173] Fascellaro G, Petrera A, Lai ZW, *et al.* Comprehensive Proteomic Analysis of Nitrogen-Starved *Mycobacterium smegmatis* Δpup Reveals the Impact of Pupylation on Nitrogen Stress Response. J Proteome Res 2016; 15(8): 2812-25.
[http://dx.doi.org/10.1021/acs.jproteome.6b00378] [PMID: 27378031]

[174] Ansong C, Wu S, Meng D, *et al.* Top-down proteomics reveals a unique protein S-thiolation switch in *Salmonella Typhimurium* in response to infection-like conditions. Proc Natl Acad Sci USA 2013; 110(25): 10153-8.
[http://dx.doi.org/10.1073/pnas.1221210110] [PMID: 23720318]

[175] Kelleher NL. Peer reviewed: Top-down proteomics. ACS Publications 2004.

[176]   Guest PC, Gottschalk MG, Bahn S. Proteomics: improving biomarker translation to modern medicine?. Springer 2013.

[177]   Freedman JC, Li J, Uzal FA, McClane BA. Proteolytic processing and activation of Clostridium perfringens epsilon toxin by caprine small intestinal contents. MBio 2014; 5(5): e01994-14.
[http://dx.doi.org/10.1128/mBio.01994-14] [PMID: 25336460]

[178]   Deracinois B, Flahaut C, Duban-Deweer S, Karamanos Y. Comparative and quantitative global proteomics approaches: an overview. Proteomes 2013; 1(3): 180-218.
[http://dx.doi.org/10.3390/proteomes1030180] [PMID: 28250403]

[179]   Marko-Varga G, Fehniger TE. Proteomics and disease--the challenges for technology and discovery. J Proteome Res 2004; 3(2): 167-78.
[http://dx.doi.org/10.1021/pr049958+] [PMID: 15113092]

[180]   López-Maury L, Marguerat S, Bähler J. Tuning gene expression to changing environments: from rapid responses to evolutionary adaptation. Nat Rev Genet 2008; 9(8): 583-93.
[http://dx.doi.org/10.1038/nrg2398] [PMID: 18591982]

[181]   Pisithkul T, Patel NM, Amador-Noguez D. Post-translational modifications as key regulators of bacterial metabolic fluxes. Curr Opin Microbiol 2015; 24: 29-37.
[http://dx.doi.org/10.1016/j.mib.2014.12.006] [PMID: 25597444]

[182]   Brunk E, Chang RL, Xia J, *et al.* Characterizing posttranslational modifications in prokaryotic metabolism using a multiscale workflow. Proc Natl Acad Sci USA 2018; 115(43): 11096-101.
[http://dx.doi.org/10.1073/pnas.1811971115] [PMID: 30301795]

[183]   Fuhrer T, Zamboni N. High-throughput discovery metabolomics. Curr Opin Biotechnol 2015; 31: 73-8.
[http://dx.doi.org/10.1016/j.copbio.2014.08.006] [PMID: 25197792]

[184]   Zhang Z, Wu S, Stenoien DL, Paša-Tolić L. High-throughput proteomics. Annu Rev Anal Chem (Palo Alto, Calif) 2014; 7: 427-54.
[http://dx.doi.org/10.1146/annurev-anchem-071213-020216] [PMID: 25014346]

[185]   Berger B, Peng J, Singh M. Computational solutions for omics data. Nat Rev Genet 2013; 14(5): 333-46.
[http://dx.doi.org/10.1038/nrg3433] [PMID: 23594911]

[186]   Baeza J, Dowell JA, Smallegan MJ, *et al.* Stoichiometry of site-specific lysine acetylation in an entire proteome. J Biol Chem 2014; 289(31): 21326-38.
[http://dx.doi.org/10.1074/jbc.M114.581843] [PMID: 24917678]

[187]   Becker D, Selbach M, Rollenhagen C, *et al.* Robust Salmonella metabolism limits possibilities for new antimicrobials. Nature 2006; 440(7082): 303-7.
[http://dx.doi.org/10.1038/nature04616] [PMID: 16541065]

[188]   Fernández-Arenas E, Cabezón V, Bermejo C, *et al.* Integrated proteomics and genomics strategies bring new insight into Candida albicans response upon macrophage interaction. Mol Cell Proteomics 2007; 6(3): 460-78.
[http://dx.doi.org/10.1074/mcp.M600210-MCP200] [PMID: 17164403]

[189]   Twine SM, Mykytczuk NC, Petit MD, *et al. in vivo* proteomic analysis of the intracellular bacterial pathogen, Francisella tularensis, isolated from mouse spleen. Biochem Biophys Res Commun 2006; 345(4): 1621-33.
[http://dx.doi.org/10.1016/j.bbrc.2006.05.070] [PMID: 16730660]

[190]   Rogers JB, DuTeau NM, Reardon KF. Use of 16S-rRNA to investigate microbial population dynamics during biodegradation of toluene and phenol by a binary culture. Biotechnol Bioeng 2000; 70(4): 436-45.
[http://dx.doi.org/10.1002/1097-0290(20001120)70:4<436::AID-BIT9>3.0.CO;2-6]   [PMID: 11005926]

[191] Higuchi R, Fockler C, Dollinger G, Watson R. Kinetic PCR analysis: real-time monitoring of DNA amplification reactions. Biotechnology (N Y) 1993; 11(9): 1026-30.
[PMID: 7764001]

[192] Müller S, Lösche A, Bley T, Scheper T. A flow cytometric approach for characterization and differentiation of bacteria during microbial processes. Appl Microbiol Biotechnol 1995; 43(1): 93-101.
[http://dx.doi.org/10.1007/BF00170629]

[193] Luo R, Fang L, Jin H, *et al.* Label-free quantitative phosphoproteomic analysis reveals differentially regulated proteins and pathway in PRRSV-infected pulmonary alveolar macrophages. J Proteome Res 2014; 13(3): 1270-80.
[http://dx.doi.org/10.1021/pr400852d] [PMID: 24533505]

[194] Shui W, Gilmore SA, Sheu L, Liu J, Keasling JD, Bertozzi CR. Quantitative proteomic profiling of host-pathogen interactions: the macrophage response to *Mycobacterium tuberculosis* lipids. J Proteome Res 2009; 8(1): 282-9.
[http://dx.doi.org/10.1021/pr800422e] [PMID: 19053526]

[195] Benndorf D, Balcke GU, Harms H, von Bergen M. Functional metaproteome analysis of protein extracts from contaminated soil and groundwater. ISME J 2007; 1(3): 224-34.
[http://dx.doi.org/10.1038/ismej.2007.39] [PMID: 18043633]

[196] Mawuenyega KG, Forst CV, Dobos KM, *et al. Mycobacterium tuberculosis* functional network analysis by global subcellular protein profiling. Mol Biol Cell 2005; 16(1): 396-404.
[http://dx.doi.org/10.1091/mbc.e04-04-0329] [PMID: 15525680]

[197] Wilmes P, Bond PL. The application of two-dimensional polyacrylamide gel electrophoresis and downstream analyses to a mixed community of prokaryotic microorganisms. Environ Microbiol 2004; 6(9): 911-20.
[http://dx.doi.org/10.1111/j.1462-2920.2004.00687.x] [PMID: 15305916]

[198] Lange V, Malmström JA, Didion J, *et al.* Targeted quantitative analysis of Streptococcus pyogenes virulence factors by multiple reaction monitoring. Mol Cell Proteomics 2008; 7(8): 1489-500.
[http://dx.doi.org/10.1074/mcp.M800032-MCP200] [PMID: 18408245]

[199] Karlsson C, Malmström L, Aebersold R, Malmström J. Proteome-wide selected reaction monitoring assays for the human pathogen Streptococcus pyogenes. Nat Commun 2012; 3(1): 1301.
[http://dx.doi.org/10.1038/ncomms2297] [PMID: 23250431]

[200] Wilmes P, Bond PL. Microbial community proteomics: elucidating the catalysts and metabolic mechanisms that drive the Earth's biogeochemical cycles. Curr Opin Microbiol 2009; 12(3): 310-7.
[http://dx.doi.org/10.1016/j.mib.2009.03.004] [PMID: 19414280]

CHAPTER 6

Pupylation: A Novel Proteolysis Pathway in Prokaryotes Functionally Reminiscent to Eukaryotic Ubiquitination

Yogesh K. Dhuriya[1] and **Divakar Sharma**[2,*]

[1] *Developmental Toxicology Laboratory, Systems Toxicology and Health Risk Assessment Group, CSIR-Indian Institute of Toxicology Research (CSIR-IITR), Vishvigyan Bhawan; 31, Mahatma Gandhi Marg Lucknow – 226 001, India*

[2] *CRF, Mass Spectrometry Laboratory, Kusuma School of Biological Sciences (KSBS), Indian Institute of Technology, Delhi (IIT-D)110016, India*

Abstract: Posttranslational modification of proteins is a prevalent method for the regulation of cells according to changes in the surrounding environment and diversifications. Pupylation, architecturally similar but not homologous to eukaryotic proteasomal degradation machinery, exists in a certain order of bacteria, especially actinobacteria. Pupylation supports the bacteria to survive under challenging environmental conditions like stress (physical or chemical) and nutrient starvation. Pupylation is also involved in iron homeostasis, which is necessary for cellular metabolism and the normal growth of bacteria. Pupylation is a posttranslational modification through which intrinsically disordered proteins are tagged for proteasomal degradation. Although this process is functionally reminiscent of ubiquitination in eukaryotes; it is carried out by a different set of enzymes in evolutionarily connected bacterial carboxylate-amine ligases. In this chapter, we will discuss the recent advances in the understanding of how proteins are tagged for proteasomal degradation in actinobacteria and its role in the survival of mycobacterium during pathogenesis in the host. Furthermore, we will examine the role of accessory factors associated with the proteasomal system in bacteria that function independently of proteolysis.

Keywords: Mpa (Mycobacterial proteasomal ATPase) and Dop (Deamidase of Pup), Mycobacterium, Proteasome, Pup, Pupylation.

INTRODUCTION

Regulated proteolysis is a fundamental process involved in posttranslational regulation especially to environmental changes, intracellular stress, and removal

* **Corresponding author Divakar Sharma:** CRF, Mass Spectrometry Laboratory, Kusuma School of Biological Sciences (KSBS), Indian Institute of Technology, Delhi (IIT-D)110016, India; Tel: +91-7906026680; E-mail: divakarsharma88@gmail.com

Divakar Sharma (Ed.)
All rights reserved-© 2020 Bentham Science Publishers

of disordered proteins. Removal and addition of functional groups on proteins increase the structural and functional diversification of proteins; these modifications alter the function, stability, localization, function, and regulation of proteins.The modification that targets the proteins towards proteasomal degradation affects the stability of proteins and has been extensively studied amongst protein modifications. Bacteria depend exclusively upon compartmentalized protease complex to degrade the proteins as they lack the sorting process in contrast to eukaryotes. In bacteria, the classical protease complex is Lon, Clp and the membrane attached Ftsh protease complex, the homologs of these proteases are also found in mitochondria and chloroplast. In addition to this, Mycobacteria and some other actinobacteria possess proteasome [1 - 3] which is not found in other bacteria. Interestingly, it is not essential under the mycobacterium culture condition [2, 4] though it persists in mycobacteria suggesting its important role in the survival of mycobacteria in the host in a specific environment [1, 2]. Pupylation, a posttranslational modification firstly observed in *Mycobacterium tuberculosis* and *Mycobacterium smegmatis* [5], suggests that prokaryotes also employ the macromolecular tags. Several lines of evidence demonstrate that modification of bacterial proteins with Pup occurs by a different pathway in contrast to eukaryotic ubiquitination [6]. Pupylation involves two homologous sequential events that involve different enzymology, Dop (deamidase of Pup), deamidation of C-terminal glutamine residue on Pup into glutamate being the first step, and then it attaches to target proteins through PafA (Proteasome accessory factorA). The various roles of pupylation have been identified for bacterial physiology is the degradation of a pupylated substrate; one of the most notable roles essential for virulence of *Mycobacterium tuberculosis*. Pupylation renders the proteins to proteasomal degradation, but not all Pup-conjugated proteins undergo this fate [7, 8]. Pupylation also regulates the activity of Mpa (Mycobacterial proteasomal ATPase) by rendering it functionally inactive [8, 9]. Further existence of depupylation in actinobacteria suggests broader role of pupylation in the bacteria [5, 10], showing that the signaling behind this process in future may help in a better understanding of the role of pupylation in the bacteria.

Bacterial Proteasome: An Evolutionary Precursor of Eukaryotic Proteasome

The proteasome is found in all three domains of living organisms that carry out the removal of damaged, unfolded and non-functional proteins. The existence of 20S CP (20S core particle) proteasome is exclusively found in actinobacteria, adopted by horizontal gene transfer during evolution [2, 4, 11]. Genes that encode this proteasome complex were not observed in standard culture conditions; hence evolution selected this proteasome for a specific condition [2, 12]. The first indication of the presence of bacterial proteasome comes from the study of

Frankia [13]. Genes (prcA and prcB) encoded to the proteasomal subunits were characterized by *Frankia, Steptomyces coelicolor* and *Rhodococcus erythropolis* [14 - 16]. The proteasomal degradation system in mycobacteria was firstly investigated in *Mycobacterium smegmatis* [4]. Lack of degradation activity in *Msm* ΔpcrB strain has been observed in *Mycobacterium smegmatis*, which suggests the presence of functional proteasome in mycobacterium. Further, the identification of prcA and prcB in *Mycobacterium tuberculosis* reveals that the proteome supports this notion [17]. Bacterial proteasome, the heptameric barrel-shaped structure, contains two rings of homo-heptameric of α (prcA) and two rings of β (prcB) subunits [3, 18, 19]. As similar to eukaryotes, the presence of threonine nucleophile in the β subunit of the bacterial proteasome is the cause of catalytic activity [20]. Gene *mpa* encodes an ATPase (equal to Regulatory particle of eukaryotic proteasome) found to co-localize with proteasome encoding gene (CP) in actinobacteria [21]. *Mycobacterium tuberculosis* lacking *mpa* gene is sensitive to nitrosative and oxidative stress. Similar phenotypes have been observed by the addition of proteasome inhibitor in mycobacterium culture [22]. In addition to this, a mutation in the mpa gene exhibits an altered level of some proteins in contrast to wild type strain, which provides evidence that the product of mpa gene plays a role in the proteasome degradation pathway [23]. ARC (AAA-ATPase ring-shaped complex) and Mpa perform a similar function as AAA-ATPase of a regulatory particle (RP) of eukaryotic proteasome while strong interaction between CPs and these proteins has not been observed [21, 24], suggesting that either it transiently interacts with these proteins or requires an accessory factor or co-factors.

Structure – *Mycobacterium* Proteasome

Structure of the *Mycobacterium* proteasome is just like archaeal CP and eukaryotic proteasome [2, 25], consisting of four stacked barrel-shaped structures, which is made up of two central β-rings flanked by alpha on both sides ($\alpha_7\beta_7$ β_7 α_7). *Mycobacterium* proteasome shows only 32% sequence identity with archaeal CP of *Thermoplasma* while 65% with bacterial CP of *Rhodococcus*. Despite this difference, the 3D structure of CP of these bacteria is superimposable with each other [26 - 28]. The proteolytic active sites have been found to associate with β-subunit of CPs [29], which synthesize as the N-terminal peptide; on auto-cleavage, it exposes the threonine acting as a nucleophile [30]. The proteolytic activity of *Mycobacterium* CP is somewhat different from the CPs of other actinobacterial species as Mycobacterium CP, which exhibits caspase and tryptic catalytic activity in contrast to archaeal CPs [2, 26]. Structural analysis of *Mycobacterium* CP disclosed that all proteolytic activity is performed by a single type of β subunit in contrast to eukaryotic CP [2, 26]. Alpha rings of bacterial CP form the pores on either end of central β-rings to prevent the undesired protein degradation. In eukaryotic CP, N-termini of alpha subunits interact with each other through YD motifs (conserved tyrosine and aspartate residue) sealing the entrance of CP, a mutation in the 7 N-terminal amino acid residues of alpha subunits resulting in disruption of gate sealing with altered peptidase activity [31]. Experimental studies have demonstrated that *Mycobacterium* CP is also a gated complex lacking the YD motif [18, 32]. Both bacterial CP and eukaryotic alpha subunit possess additional N-termini helix termed as H0 (Homologous to β subunit) which protrudes radially outwards from the central pore of α-ring [29, 32, 33]. N-termini of alpha subunit is linked to H0 either directly (bacterial CP) or through reverse turn element in eukaryotic CP leading to the closure of the gate.

Mpa –*Mycobacterium* Proteasome ATPase

ATP hydrolyzing interactors process the naïve proteins before entering the inner chamber of 20S CP [34]. These interactors unfold the tagged proteins by using ATP and translocate them into the inner chamber of proteolytic machinery where degradation takes place [2, 35]. 19S RP of eukaryotic proteasome possesses AAA-type hexameric ATPase subunits and several non-ATPase subunits [36]. Similarly, *Mycobacterium* CP interacts with AAA type ATPase termed as Mpa (Mycobacterial proteasomal ATPase), while the C-terminal AAA modules form the ring structure which stacks on the CP through the C-terminal GQYL motif [24]. Two oligonucleotides domain (OD) comes after AAA domain in Mpa, adopting five stranded structures resembling β-barrel fold. From 21-58 amino acid residues of Pup (of 64 residues) participate in the helix formation when Mpa interacts with Pup and is arranged into three helical coiled-coil structures in antiparallel fashion with N-termini of Mpa [37, 38]. N-termini of Pup protein remains unstructured, responsible for the initiation of the unfolding of proteins after binding to the translocation loop of Mpa [2, 39]. Energy-dependent upward and downward movement of tagged substrates leads to the unfolding of damaged proteins while the translocation initiation model is not mediated with ubiquitin and ubiquitinated substrates, which must lose C or N terminal or extended loop to reach inside the proteasomal core [40, 41]. *In vitro* studies have demonstrated that Pup acts as a threading handle and degrades along with substrate inside the proteasome core in contrast to ubiquitin [39]. Future studies may reveal the additional molecular mechanisms which prevent the *in-vivo* co-degradation of tagged substrates. A few studies hypothesized that Depupylase may be involved in this process [42] while there are no clear pieces of evidence available for the association of depupylase with *in vivo* co-degradation.

Pupylation –Mark of Intrinsic Protein Demolition

Posttranslational modification of proteins occurs before it undergoes proteasomal degradation. In eukaryotes, protein tagging is mediated by ubiquitin, while Pup is responsible for proteins marking in bacteria; its gene is located upstream of the CP gene in proteasome containing bacteria. The addition of Pup on unfolded or damaged proteins is considered as pupylation (Fig. **1**) and helps in the identification of pupylated targets at the ATPase Mpa [43]. However, ubiquitination and pupylation are biochemically different processes as they evolutionary conserve to perform similar functions. Pup possesses a diglycine motif at the terminal position of the C-terminus followed by either glutamine or glutamate residues depending on the organism. Diglycine motifs at the C-terminus as well as its small size are the common features shared by Pup and ubiquitin proteins. Initially, the researcher thought that the removal of glutamate or glycine from the C-terminal for conjugation would be analogous to the ubiquitination process [44]. Mass spectrometry analysis on *Mycobacterium tuberculosis* and *Mycobacterium smegmatis* has shown that glutamine is not removed from the C-terminus of Pup, instead deamidation of glutamine residue into glutamate before interacting to a lysine residue on target proteins [43, 45]. Deamidation reaction is mediated by Dop, encoded by a gene that is present just upstream of Pup and CP genes [6, 35]. Bioinformatics analysis has shown that structurally Dop is homologous to carboxylate-amine/ammonia ligase belonging to glutamine synthetases family, not to ubiquitin ligase enzymes [46]. In spite of this, PafA also shows structural homology to carboxylate-amine/ammonia ligase superfamily observed through bioinformatic analysis [46]. In the presence of ATP and Dop, PafA mediates the conjugation reaction by linking Pup to proteasome substrates PanB and FabD during the *in-vitro* condition [35]. ATP is needed for conjugation reaction mediated by PafA, suggesting that it involves phosphorylated intermediate. The Pup with terminal glutamate residue (Pup-Glu), not with glutamine (Pup-Gln), is a substrate for PafA mediated conjugation reaction without the involvement of Dop, suggesting that deamidation is followed by conjugation. A single enzyme is needed for deamidation and conjugation reaction in bacteria where pup gene encodes Pup with terminal glutamate residue. In *Mycobacterium tuberculosis*, a genetic mutation in pafA disrupts the pupylation process [43], suggesting that PafA mediates most of the pupylation reaction but not all pupylation reactions during the *in-vivo* condition, which explains the existence of another type of PafA in bacteria or involvement of accessory factors that alter the activity of PafA leading to the different destination of the pupylated substate instead of proteasome targeting. Experimental studies have demonstrated that Pup did not make polymeric chains on the targeted substrate like ubiquitin [35, 43, 45]. It is still to be investigated whether a single substrate has multiple pupylated lysines or not and what are the consequences of these multiple pupylated lysines in bacteria.

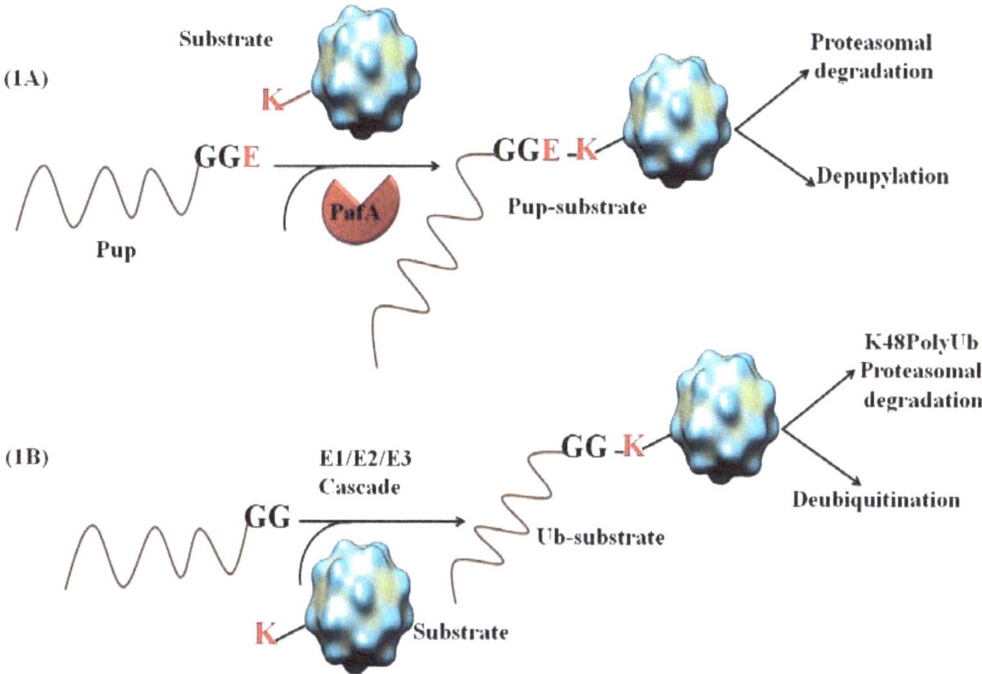

Fig. (1). Addition of Pup protein **(1A)** on the damaged or unfolded protein is referred to as pupylation resulting in proteasomal degradation. In eukaryotes, proteasomal degradation occurs through ubiquitination **(1B)**. In pupylation, substrate lysine is linked through the C-terminal glutamate of Pup, while ubiquitin is linked through its C-terminal diglycine motif.

Reversed Pupylation –Depupylation

Mycobacterial cells can reverse the pupylation process by breaking the isopeptide bond in the pupylated substrate [2, 19] termed as depupylation (Fig. **2**). Depupylase (Dop), a structural homologue of PafA acts as an antagonist of pupylation by removing Pup from tagged substrates and echoes with deubiquitinating enzymes of eukaryotes. The evolutionary origin of Dop is similar to PafA as both require a surface to interact with Pup and target substrate, both mediate the nucleophilic attack on the carbonyl carbon of lysine residue or water molecule (Dop) acting as a nucleophile. Dop and PafA possess a conserved aspartate residue, which plays a supporting role in nucleophile attack on carbonyl carbon [47]. Aspartate residue of Dop plays a supporting role during the nucleophile attack leading to the formation of transient covalent bonds with Pup [48]. *In vivo* assessment of depupylation activity is very difficult because Dop also mediates the deamidation reaction. However, many actinobacteria express the

Pup with glutamate instead of glutamine, which exhibits the depupylation activity as bypassing of the deamidation step [2]. An altered level of pupylated substrate has been found in the *dop* mutant of *Mycobacterium tuberculosis* in contrast to wild type strain [42]. The above studies suggested that depupylation is necessary for the recycling of pups to prevent the proteasomal degradation of doomed proteins.

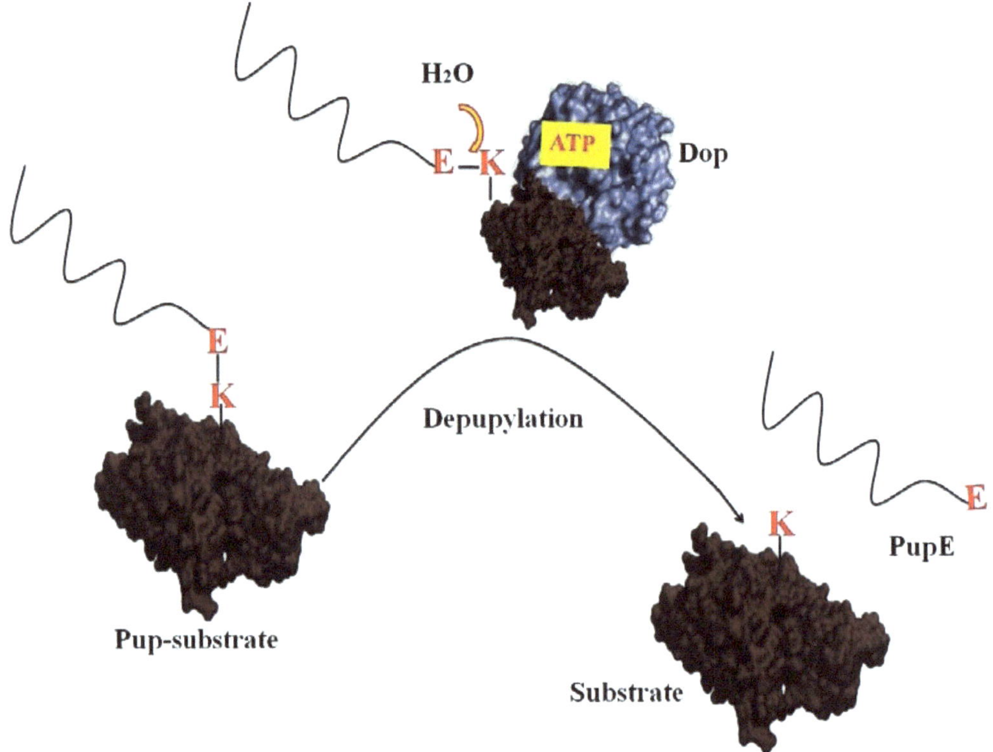

Fig. (2). Depupylation, a process to reverse the pupylation by breaking the isopeptide bond of pupylated substrates which is mediated by depupylase (Dop) enzyme. Dop employs water molecules as a nucleophile.

Recycling of Pup – Proteasome Regulation

The survival of organisms depends upon the ability to modify theirphenotypes according to the surrounding environment, which requires tightly controlled synthesis and degradation of specific proteins. Due to this, protein degradation must be regulated in a precise manner to avoid the removal of the undesired one. A study led by Cerda-Maira on Δ*dop* mutants; reported the importance of Pup recycling in *Mycobacterium,* which remains controversial due to contradicting results in *Mycobacterium smegmatis* [2, 42]. Recently, a study settled this contradiction by keeping the slow rate of Pup synthesis to maintain the stable pupylome in the absence of Pup recycling in *Mycobacterium smegmatis* [49, 50]. Different types of regulation exist for the proteolytic process as they exhibit a broad spectrum of substrate specificity; PafA marks the several proteins for demolition through proteasome; therefore, the low concentration of Pup limits the activity of PafA to tag the proteins [49]. Pup recycling is regulated by Dop, Pup, and proteasome to maintain the pupylome for PafA drove pupylation, hence the steady-state of pupylation and degradation is easy to maintain through Pup-proteasome system (PPS) and can be easily tuned by changing the concentration of Pup in cytoplasm and proteasome function (Fig. **3**). Under the starvation condition, both factors have been found to be upregulated, leading to increased pupylome level and enhanced pupylation-degradation cycles [51]. However, still much evidence is required to unveil the precise molecular mechanism behind the Pup recycling and associated pup-proteasome system.

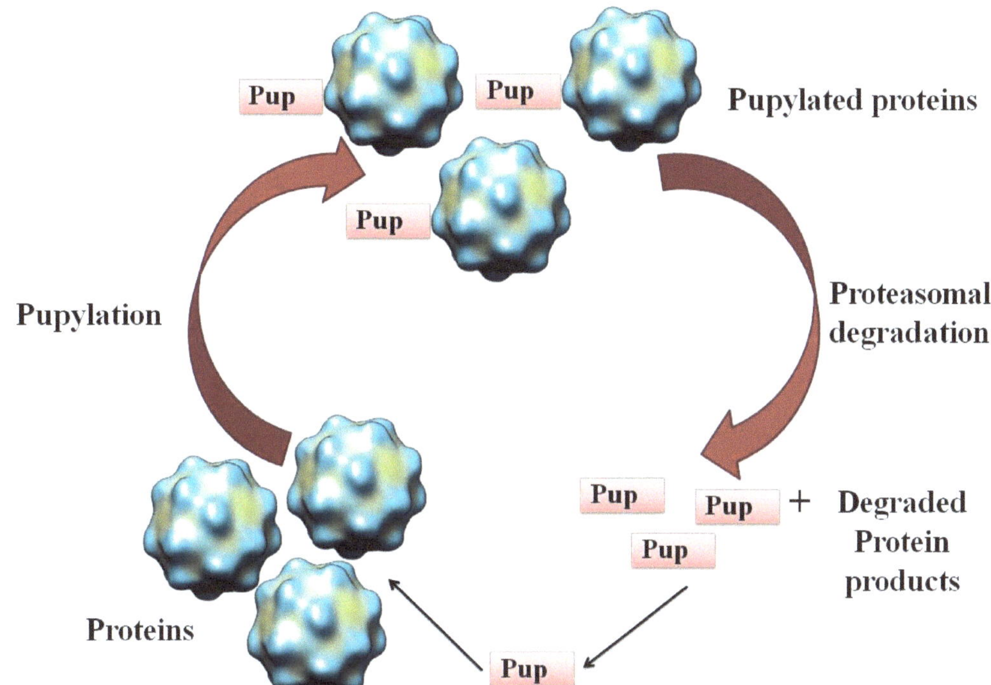

Fig. (3). Recycling of Pup protein is essential as a low level of Pup limits the PafA driven pupylation of protein. Steady-state of pupylation and degradation is achieved through Pup-proteasome system (PPS) and can be changed by altering the level of Pup and the function of proteasome.

Pup-proteasome System (PPS) – Bacterial Physiology

In vivo experiments reported that pup driven protein degradation is involved in various functions. The silencing of the proteasome in *Mycobacterium tuberculosis*-infected murine model decreased the bacterial counts, while a mutation in *pafA* gene led to the increased survival rate of infected mice [3, 52]. Pup-proteasome complex reduced the toxic effect of NO (nitric oxide) generated by macrophages leading to persistence of *Mycobacterium tuberculosis* inside macrophage [22, 53]. Proteolytic products of cytokinin showed bacteriotoxic effects with NO generated by *Mycobacterium tuberculosis*, requiring the instant removal of cytokinin producing mycobacterium enzymes by the pup-proteasome system. Altered pupylation impairs the survival of *Mycobacterium smegmatis* in the nitrogen-deficient medium [51], suggesting that the pup-proteasome system might be involved in amino acid recycling to furnish the precursors during nitrogenous biosynthesis precursors. Further, proteomic analysis on Msm Δpup (Pup deleted strain) has shown the direct connection of pupylation with the assimilation of nitrogen in *Mycobacterium smegmatis* [54]. In spite of this, pup-proteasome is also involved in iron homeostasis by targeting the ferritin protein, which mediates the iron storage in mycobacteria [8]. Several lines of evidence have reported the role of the pup-proteasome system in mycobacterium DNA damage response [55, 56]. PafBC (a protein complex) plays a key role during the DNA damage response in mycobacterium and activates the transcription of many genes involved in DNA damage and oxidative stress. A recent study has shown that it alters the plasmid copy number, suggesting that transcription of *pafA* gene is controlled by PafB [57]. The above studies suggest that pup-proteasom--mediated protein degradation maintains the stress response by the removal of upregulated proteins during DNA damage and oxidative stress.

Proteasome – Stress Sensor Machine

Pup-proteasome dependent protein degradation occurs by using ATP; however, energy independent activators of the proteasome (PA complexes) have also been observed in eukaryotes, including PA200 [58] and PA28 [59], leading to the formation of a PA-20S CP complex. Recent studies discovered the energy-independent ring-shaped proteasome activators in bacteria termed as Bpa (bacterial proteasomal activators), which forms the complex with 20S CP [60, 61]. Bpa employs a similar motif (GxYx) as Mpa to enter binding pockets found between the alpha subunits of 20S CP. Salt bridge formation has been observed between the C-terminal carboxylate of Bpa and lysine residue of the proteasomal binding pocket and penultimate tyrosine residue with arginine in the binding pockets of the proteasome [62]. Bpa possesses a single domain with a four-helix bundle that lacks intrinsic ATP hydrolyzing activity in contrast to Mpa. Promoters of Bpa form a twelve-membered ring structure creating a wide platform for substrate binding [62, 63]. Bpa-proteasome complex mediates degradation of the unstructured model substrate, suggesting that it removes the non-native and damaged proteins under stress conditions. HspR (heat shock repressor) was recognized as the first degradation substrate for Bpa mediated degradation pathway [61]. HspR down-regulates the expression of ClpB and Hsp70; these two chaperones play a key role in protein quality control [64]. The Bpa knockout bacterial strain exhibits a heat-sensitive phenotype [61], suggesting the importance of the Bpa mediated degradation pathway under stress.

Cpa (Cdc48-like protein of actinobacteria) –A Novel Proteasome Interacting Molecule

Endoplasmic resident proteins undergo degradation *via* ERAD (endoplasmic reticulum-associated degradation) pathway, which requires translocation of ER-resident proteins fromER to the cytoplasm [65]. In the ERAD pathway, proteasome forms the complex with Cdc48 protein [66]. Several lines of evidence have demonstrated that Cdc48 interacts with 20S proteasome resulting in the formation of Cdc48-proteasome complex [67 - 69]. Many actinobacteria and mycobacteria possess a homolog of Cdc48 termed as Cpa [70, 71]. A recent study has shown that Cpa forms the Cpa-proteasome complex in bacteria after interacting with 20S CP [71]. However, C-termini of Cpa do not possess penultimate tyrosine residue in contrast to Mpa and Bpa, which mediates interaction with proteasomal pockets. A knockout *cpa* strain of *Mycobacterium smegmatis* shows growth defects under carbon malnourishment and expression of several proteins which has been found to increase in contrast to control [71]. Several recent studies observed accumulation and depletion, while no conclusion has been reported about potential substrates of Cpa. Further studies are needed to identify the exact substrate for Cpa mediated degradation pathway.

CONCLUSION AND FUTURE PROSPECTS

The pupylation machinery in mycobacteria and some other actinobacteria proved the existence of regulated proteolysis, which is essential to cop-up with the surrounding environment. It is functionally analogous to eukaryotic machinery while having different evolutionary origins and different modes of protein modification. *Mycobacterium tuberculosis* employs this machinery to overcome the defense system of the host, unraveling the molecular mechanisms behind this rescue that may help us understand the virulence process inside the host and drug resistance [72 - 76]. Depupylation helps to maintain the pupylome level by recycling the Pup protein inside the bacteria. However, still considerable evidence is required to unravel the precise molecular mechanism behind the Pup recycling and associated pup-proteasome system. Either reversed horizontal gene transfer (from eukaryotes to mycobacteria) or horizontal gene transfer is the main reason behind the evolution of the Pup-proteasome system in bacteria, and is still the point of debate. Understanding the proteasome system in a better way may help in the development of antibacterial drugs, especially against *Mycobacterium tuberculosis*, a causative agent of the deadliest disease in humans. Both ATP dependent and independent protein degradation pathway exist in bacteria and further evidence is needed to know that these two pathways interact somewhere during degradation to perform a distinct function. Further studies may reveal other various cellular functions that are mediated by the pup-proteasome system, in future.

CONSENT FOR PUBLICATION

Not applicable.

CONFLICT OF INTEREST

The authors declare no conflict of interest, financial or otherwise.

ACKNOWLEDGEMENTS

Declared none.

REFERENCES

[1]   Knipfer N, Seth A, Roudiak SG, Shrader TE. Species variation in ATP-dependent protein degradation: protease profiles differ between mycobacteria and protease functions differ between *Mycobacterium smegmatis* and *Escherichia coli*. Gene 1999; 231(1-2): 95-104.
[http://dx.doi.org/10.1016/S0378-1119(99)00087-6] [PMID: 10231573]

[2]   Imkamp F, Ziemski M, Weber-Ban E. Pupylation-dependent and -independent proteasomal degradation in mycobacteria. Biomol Concepts 2015; 6(4): 285-301.
[http://dx.doi.org/10.1515/bmc-2015-0017] [PMID: 26352358]

[3]   Müller AU, Weber-Ban E. The bacterial proteasome at the core of diverse degradation pathways. Front Mol Biosci 2019; 6: 23.
[http://dx.doi.org/10.3389/fmolb.2019.00023] [PMID: 31024929]

[4]   Knipfer N, Shrader TE. Inactivation of the 20S proteasome in *Mycobacterium smegmatis*. Mol Microbiol 1997; 25(2): 375-83.
[http://dx.doi.org/10.1046/j.1365-2958.1997.4721837.x] [PMID: 9282749]

[5]   Barandun J, Delley CL, Weber-Ban E. The pupylation pathway and its role in mycobacteria. BMC Biol 2012; 10(1): 95.
[http://dx.doi.org/10.1186/1741-7007-10-95] [PMID: 23198822]

[6]   Striebel F, Imkamp F, Özcelik D, Weber-Ban E. Pupylation as a signal for proteasomal degradation in bacteria. Biochimica et Biophysica Acta (BBA)-. Molecular Cell Research 2014; 1843(1): 103-13.

[7]   Boubakri H, Seghezzi N, Duchateau M, *et al.* The absence of pupylation (prokaryotic ubiquitin-like protein modification) affects morphological and physiological differentiation in Streptomyces coelicolor. J Bacteriol 2015; 197(21): 3388-99.
[http://dx.doi.org/10.1128/JB.00591-15] [PMID: 26283768]

[8]   Küberl A, Polen T, Bott M. The pupylation machinery is involved in iron homeostasis by targeting the iron storage protein ferritin. Proc Natl Acad Sci USA 2016; 113(17): 4806-11.
[http://dx.doi.org/10.1073/pnas.1514529113] [PMID: 27078093]

[9]   Delley CL, Striebel F, Heydenreich FM, Özcelik D, Weber-Ban E. Activity of the mycobacterial proteasomal ATPase Mpa is reversibly regulated by pupylation. J Biol Chem 2012; 287(11): 7907-14. [http://dx.doi.org/10.1074/jbc.M111.331124] [PMID: 22210775]

[10]   Burns KE, Cerda-Maira FA, Wang T, Li H, Bishai WR, Darwin KH. "Depupylation" of prokaryotic ubiquitin-like protein from mycobacterial proteasome substrates. Mol Cell 2010; 39(5): 821-7. [http://dx.doi.org/10.1016/j.molcel.2010.07.019] [PMID: 20705495]

[11]   De Mot R, Schoofs G, Nagy I. Proteome analysis of Streptomyces coelicolor mutants affected in the proteasome system reveals changes in stress-responsive proteins. Arch Microbiol 2007; 188(3): 257-71. [http://dx.doi.org/10.1007/s00203-007-0243-8] [PMID: 17486317]

[12]   Powers ET, Balch WE. Diversity in the origins of proteostasis networks--a driver for protein function in evolution. Nat Rev Mol Cell Biol 2013; 14(4): 237-48. [http://dx.doi.org/10.1038/nrm3542] [PMID: 23463216]

[13]   Benoist P, Müller A, Diem HG, Schwencke J. High-molecular-mass multicatalytic proteinase complexes produced by the nitrogen-fixing actinomycete Frankia strain BR. J Bacteriol 1992; 174(5): 1495-504. [http://dx.doi.org/10.1128/JB.174.5.1495-1504.1992] [PMID: 1537794]

[14]   Tamura T, Nagy I, Lupas A, *et al.* The first characterization of a eubacterial proteasome: the 20S complex of Rhodococcus. Curr Biol 1995; 5(7): 766-74. [http://dx.doi.org/10.1016/S0960-9822(95)00153-9] [PMID: 7583123]

[15]   Nagy I, Tamura T, Vanderleyden J, Baumeister W, De Mot R. The 20S proteasome of Streptomyces coelicolor. J Bacteriol 1998; 180(20): 5448-53. [http://dx.doi.org/10.1128/JB.180.20.5448-5453.1998] [PMID: 9765579]

[16]   Pouch MN, Cournoyer B, Baumeister W. Characterization of the 20S proteasome from the actinomycete Frankia. Mol Microbiol 2000; 35(2): 368-77. [http://dx.doi.org/10.1046/j.1365-2958.2000.01703.x] [PMID: 10652097]

[17]   Mollenkopf HJ, Jungblut PR, Raupach B, *et al.* A dynamic two-dimensional polyacrylamide gel electrophoresis database: the mycobacterial proteome *via* Internet. Electrophoresis 1999; 20(11): 2172-80. [http://dx.doi.org/10.1002/(SICI)1522-2683(19990801)20:11<2172::AID-ELPS2172>3.0.CO;2-M] [PMID: 10493122]

[18]   Lin G, Hu G, Tsu C, *et al. Mycobacterium tuberculosis* prcBA genes encode a gated proteasome with broad oligopeptide specificity. Mol Microbiol 2006; 59(5): 1405-16. [http://dx.doi.org/10.1111/j.1365-2958.2005.05035.x] [PMID: 16468985]

[19]   Burns KE, Darwin KH. Pupylation *versus* ubiquitylation: tagging for proteasome-dependent degradation. Cell Microbiol 2010; 12(4): 424-31. [http://dx.doi.org/10.1111/j.1462-5822.2010.01447.x] [PMID: 20109157]

[20]   Baumeister W, Walz J, Zühl F, Seemüller E. The proteasome: paradigm of a self-compartmentalizing protease. Cell 1998; 92(3): 367-80. [http://dx.doi.org/10.1016/S0092-8674(00)80929-0] [PMID: 9476896]

[21]   Wolf S, Nagy I, Lupas A, *et al.* Characterization of ARC, a divergent member of the AAA ATPase family from Rhodococcus erythropolis. J Mol Biol 1998; 277(1): 13-25. [http://dx.doi.org/10.1006/jmbi.1997.1589] [PMID: 9514743]

[22]   Darwin KH, Ehrt S, Gutierrez-Ramos JC, Weich N, Nathan CF. The proteasome of *Mycobacterium tuberculosis* is required for resistance to nitric oxide. Science 2003; 302(5652): 1963-6. [http://dx.doi.org/10.1126/science.1091176] [PMID: 14671303]

[23]   Pearce MJ, Arora P, Festa RA, Butler-Wu SM, Gokhale RS, Darwin KH. Identification of substrates of the *Mycobacterium tuberculosis* proteasome. EMBO J 2006; 25(22): 5423-32.

[http://dx.doi.org/10.1038/sj.emboj.7601405] [PMID: 17082771]

[24]   Wang T, Li H, Lin G, *et al.* Structural insights on the *Mycobacterium tuberculosis* proteasomal ATPase Mpa. Structure 2009; 17(10): 1377-85.
[http://dx.doi.org/10.1016/j.str.2009.08.010] [PMID: 19836337]

[25]   Maupin-Furlow JA. Archaeal proteasomes and sampylation. Regulated Proteolysis in Microorganisms. Dordrecht: Springer 2013; pp. 297-327.
[http://dx.doi.org/10.1007/978-94-007-5940-4_11]

[26]   Hu G, Lin G, Wang M, *et al.* Structure of the *Mycobacterium tuberculosis* proteasome and mechanism of inhibition by a peptidyl boronate. Mol Microbiol 2006; 59(5): 1417-28.
[http://dx.doi.org/10.1111/j.1365-2958.2005.05036.x] [PMID: 16468986]

[27]   Witt S, Kwon YD, Sharon M, *et al.* Proteasome assembly triggers a switch required for active-site maturation. Structure 2006; 14(7): 1179-88.
[http://dx.doi.org/10.1016/j.str.2006.05.019] [PMID: 16843899]

[28]   Samanovic MI, Li H, Darwin KH. The pup-proteasome system of *Mycobacterium tuberculosis*. In Regulated proteolysis in microorganisms. Dordrecht: Springer 2013; pp. 267-95.

[29]   Groll M, Ditzel L, Löwe J, *et al.* Structure of 20S proteasome from yeast at 2.4 A resolution. Nature 1997; 386(6624): 463-71.
[http://dx.doi.org/10.1038/386463a0] [PMID: 9087403]

[30]   Seemuller E, Lupas A, Baumeister W. Autocatalytic processing of the 20S proteasome. Nature 1996; 382(6590): 468-71.
[http://dx.doi.org/10.1038/382468a0] [PMID: 8684489]

[31]   Groll M, Bajorek M, Köhler A, *et al.* A gated channel into the proteasome core particle. Nat Struct Biol 2000; 7(11): 1062-7.
[http://dx.doi.org/10.1038/80992] [PMID: 11062564]

[32]   Li D, Li H, Wang T, Pan H, Lin G, Li H. Structural basis for the assembly and gate closure mechanisms of the *Mycobacterium tuberculosis* 20S proteasome. EMBO J 2010; 29(12): 2037-47.
[http://dx.doi.org/10.1038/emboj.2010.95] [PMID: 20461058]

[33]   Whitby FG, Masters EI, Kramer L, *et al.* Structural basis for the activation of 20S proteasomes by 11S regulators. Nature 2000; 408(6808): 115-20.
[http://dx.doi.org/10.1038/35040607] [PMID: 11081519]

[34]   Finley D. Recognition and processing of ubiquitin-protein conjugates by the proteasome. Annu Rev Biochem 2009; 78: 477-513.
[http://dx.doi.org/10.1146/annurev.biochem.78.081507.101607] [PMID: 19489727]

[35]   Striebel F, Imkamp F, Sutter M, Steiner M, Mamedov A, Weber-Ban E. Bacterial ubiquitin-like modifier Pup is deamidated and conjugated to substrates by distinct but homologous enzymes. Nat Struct Mol Biol 2009; 16(6): 647-51.
[http://dx.doi.org/10.1038/nsmb.1597] [PMID: 19448618]

[36]   Lander GC, Estrin E, Matyskiela ME, Bashore C, Nogales E, Martin A. Complete subunit architecture of the proteasome regulatory particle. Nature 2012; 482(7384): 186-91.
[http://dx.doi.org/10.1038/nature10774] [PMID: 22237024]

[37]   Sutter M, Striebel F, Damberger FF, Allain FH, Weber-Ban E. A distinct structural region of the prokaryotic ubiquitin-like protein (Pup) is recognized by the N-terminal domain of the proteasomal ATPase Mpa. FEBS Lett 2009; 583(19): 3151-7.
[http://dx.doi.org/10.1016/j.febslet.2009.09.020] [PMID: 19761766]

[38]   Wang T, Darwin KH, Li H. Binding-induced folding of prokaryotic ubiquitin-like protein on the Mycobacterium proteasomal ATPase targets substrates for degradation. Nat Struct Mol Biol 2010; 17(11): 1352-7.
[http://dx.doi.org/10.1038/nsmb.1918] [PMID: 20953180]

[39]   Striebel F, Hunkeler M, Summer H, Weber-Ban E. The mycobacterial Mpa-proteasome unfolds and degrades pupylated substrates by engaging Pup's N-terminus. EMBO J 2010; 29(7): 1262-71.
[http://dx.doi.org/10.1038/emboj.2010.23] [PMID: 20203624]

[40]   Prakash S, Tian L, Ratliff KS, Lehotzky RE, Matouschek A. An unstructured initiation site is required for efficient proteasome-mediated degradation. Nat Struct Mol Biol 2004; 11(9): 830-7.
[http://dx.doi.org/10.1038/nsmb814] [PMID: 15311270]

[41]   Prakash S, Inobe T, Hatch AJ, Matouschek A. Substrate selection by the proteasome during degradation of protein complexes. Nat Chem Biol 2009; 5(1): 29-36.
[http://dx.doi.org/10.1038/nchembio.130] [PMID: 19029916]

[42]   Cerda-Maira FA, Pearce MJ, Fuortes M, Bishai WR, Hubbard SR, Darwin KH. Molecular analysis of the prokaryotic ubiquitin-like protein (Pup) conjugation pathway in *Mycobacterium tuberculosis*. Mol Microbiol 2010; 77(5): 1123-35.
[http://dx.doi.org/10.1111/j.1365-2958.2010.07276.x] [PMID: 20636328]

[43]   Pearce MJ, Mintseris J, Ferreyra J, Gygi SP, Darwin KH. Ubiquitin-like protein involved in the proteasome pathway of *Mycobacterium tuberculosis*. Science 2008; 322(5904): 1104-7.
[http://dx.doi.org/10.1126/science.1163885] [PMID: 18832610]

[44]   Pickart CM. Mechanisms underlying ubiquitination. Annu Rev Biochem 2001; 70(1): 503-33.
[http://dx.doi.org/10.1146/annurev.biochem.70.1.503] [PMID: 11395416]

[45]   Burns KE, Liu WT, Boshoff HI, Dorrestein PC, Barry CE III. Proteasomal protein degradation in Mycobacteria is dependent upon a prokaryotic ubiquitin-like protein. J Biol Chem 2009; 284(5): 3069-75.
[http://dx.doi.org/10.1074/jbc.M808032200] [PMID: 19028679]

[46]   Iyer LM, Burroughs AM, Aravind L. Unraveling the biochemistry and provenance of pupylation: a prokaryotic analog of ubiquitination. Biol Direct 2008; 3(1): 45.
[http://dx.doi.org/10.1186/1745-6150-3-45] [PMID: 18980670]

[47]   Özcelik D, Barandun J, Schmitz N, *et al.* Structures of Pup ligase PafA and depupylase Dop from the prokaryotic ubiquitin-like modification pathway. Nat Commun 2012; 3(1): 1014.
[http://dx.doi.org/10.1038/ncomms2009] [PMID: 22910360]

[48]   Burns KE, McAllister FE, Schwerdtfeger C, *et al. Mycobacterium tuberculosis* prokaryotic ubiquitin-like protein-deconjugating enzyme is an unusual aspartate amidase. J Biol Chem 2012; 287(44): 37522-9.
[http://dx.doi.org/10.1074/jbc.M112.384784] [PMID: 22942282]

[49]   Elharar Y, Schlussel S, Hecht N, Meijler MM, Gur E. The regulatory significance of tag recycling in the mycobacterial Pup-proteasome system. FEBS J 2017; 284(12): 1804-14.
[http://dx.doi.org/10.1111/febs.14086] [PMID: 28440944]

[50]   Petchiappan A, Chatterji D. Pup recycling regulates the proteasome. FEBS J 2017; 284(12): 1787-9.
[http://dx.doi.org/10.1111/febs.14112] [PMID: 28627115]

[51]   Elharar Y, Roth Z, Hermelin I, *et al.* Survival of mycobacteria depends on proteasome-mediated amino acid recycling under nutrient limitation. EMBO J 2014; 33(16): 1802-14.
[http://dx.doi.org/10.15252/embj.201387076] [PMID: 24986881]

[52]   Gandotra S, Schnappinger D, Monteleone M, Hillen W, Ehrt S. *In vivo* gene silencing identifies the *Mycobacterium tuberculosis* proteasome as essential for the bacteria to persist in mice. Nat Med 2007; 13(12): 1515-20.
[http://dx.doi.org/10.1038/nm1683] [PMID: 18059281]

[53]   Samanovic MI, Tu S, Novák O, *et al.* Proteasomal control of cytokinin synthesis protects *Mycobacterium tuberculosis* against nitric oxide. Mol Cell 2015; 57(6): 984-94.
[http://dx.doi.org/10.1016/j.molcel.2015.01.024] [PMID: 25728768]

[54]　Fascellaro G, Petrera A, Lai ZW, *et al.* Comprehensive proteomic analysis of nitrogen-starved *Mycobacterium smegmatis* Δpup reveals the impact of pupylation on nitrogen stress response. J Proteome Res 2016; 15(8): 2812-25.
[http://dx.doi.org/10.1021/acs.jproteome.6b00378] [PMID: 27378031]

[55]　Olivencia BF, Müller AU, Roschitzki B, Burger S, Weber-Ban E, Imkamp F. *Mycobacterium smegmatis* PafBC is involved in regulation of DNA damage response. Sci Rep 2017; 7(1): 1-3.
[PMID: 28127051]

[56]　Müller AU, Imkamp F, Weber-Ban E. The mycobacterial LexA/RecA-independent DNA damage response is controlled by PafBC and the Pup-proteasome system. Cell Rep 2018; 23(12): 3551-64.
[http://dx.doi.org/10.1016/j.celrep.2018.05.073] [PMID: 29924998]

[57]　Korman M, Elharar Y, Fishov I, Gur E. The transcription of pafA, encoding the prokaryotic ubiquitin-like protein ligase, is regulated by PafBC. Future Microbiol 2019; 14(1): 11-21.
[http://dx.doi.org/10.2217/fmb-2018-0278] [PMID: 30547686]

[58]　Savulescu AF, Glickman MH. Proteasome activator 200: the heat is on.... Mol Cell Proteomics 2011; 10(5): 006890.
[http://dx.doi.org/10.1074/mcp.R110.006890] [PMID: 21389348]

[59]　Cascio P. PA28αβ: the enigmatic magic ring of the proteasome? Biomolecules 2014; 4(2): 566-84.
[http://dx.doi.org/10.3390/biom4020566] [PMID: 24970231]

[60]　Delley CL, Laederach J, Ziemski M, Bolten M, Boehringer D, Weber-Ban E. Bacterial proteasome activator bpa (rv3780) is a novel ring-shaped interactor of the mycobacterial proteasome. PLoS One 2014; 9(12): e114348.
[http://dx.doi.org/10.1371/journal.pone.0114348] [PMID: 25469515]

[61]　Jastrab JB, Wang T, Murphy JP, *et al.* An adenosine triphosphate-independent proteasome activator contributes to the virulence of *Mycobacterium tuberculosis*. Proc Natl Acad Sci USA 2015; 112(14): E1763-72.
[http://dx.doi.org/10.1073/pnas.1423319112] [PMID: 25831519]

[62]　Bolten M, Delley CL, Leibundgut M, Boehringer D, Ban N, Weber-Ban E. Structural analysis of the bacterial proteasome activator Bpa in complex with the 20S proteasome. Structure 2016; 24(12): 2138-51.
[http://dx.doi.org/10.1016/j.str.2016.10.008] [PMID: 27839949]

[63]　Bai L, Hu K, Wang T, Jastrab JB, Darwin KH, Li H. Structural analysis of the dodecameric proteasome activator PafE in *Mycobacterium tuberculosis*. Proc Natl Acad Sci USA 2016; 113(14): E1983-92.
[http://dx.doi.org/10.1073/pnas.1512094113] [PMID: 27001842]

[64]　Bucca G, Brassington AM, Schönfeld HJ, Smith CP. The HspR regulon of Streptomyces coelicolor: a role for the DnaK chaperone as a transcriptional co-repressordagger. Mol Microbiol 2000; 38(5): 1093-103.
[http://dx.doi.org/10.1046/j.1365-2958.2000.02194.x] [PMID: 11123682]

[65]　Ruggiano A, Foresti O, Carvalho P. Quality control: ER-associated degradation: protein quality control and beyond. J Cell Biol 2014; 204(6): 869-79.
[http://dx.doi.org/10.1083/jcb.201312042] [PMID: 24637321]

[66]　Wolf DH, Stolz A. The Cdc48 machine in endoplasmic reticulum associated protein degradation. Biochim Biophys Acta 2012; 1823(1): 117-24.
[http://dx.doi.org/10.1016/j.bbamcr.2011.09.002] [PMID: 21945179]

[67]　Barthelme D, Sauer RT. Identification of the Cdc48•20S proteasome as an ancient AAA+ proteolytic machine. Science 2012; 337(6096): 843-6.
[http://dx.doi.org/10.1126/science.1224352] [PMID: 22837385]

[68]　Barthelme D, Jauregui R, Sauer RT. An ALS disease mutation in Cdc48/p97 impairs 20S proteasome

binding and proteolytic communication. Protein Sci 2015; 24(9): 1521-7.
[http://dx.doi.org/10.1002/pro.2740] [PMID: 26134898]

[69]   Esaki M, Johjima-Murata A, Islam MT, Ogura T. Biological and pathological implications of an alternative ATP-powered proteasomal assembly with Cdc48 and the 20S peptidase. Front Mol Biosci 2018; 5: 56.
[http://dx.doi.org/10.3389/fmolb.2018.00056] [PMID: 29951484]

[70]   Unciuleac MC, Smith PC, Shuman S. Crystal structure and biochemical characterization of a *Mycobacterium smegmatis* AAA-Type nucleoside triphosphatase phosphohydrolase (Msm0858). J Bacteriol 2016; 198(10): 1521-33.
[http://dx.doi.org/10.1128/JB.00905-15] [PMID: 26953339]

[71]   Ziemski M, Jomaa A, Mayer D, *et al.* Cdc48-like protein of actinobacteria (Cpa) is a novel proteasome interactor in mycobacteria and related organisms. eLife 2018; 7: e34055.
[http://dx.doi.org/10.7554/eLife.34055] [PMID: 29809155]

[72]   Sharma D, Kumar B, Lata M, *et al.* Comparative proteomic analysis of aminoglycosides resistant and susceptible *Mycobacterium tuberculosis* clinical isolates for exploring potential drug targets. PLoS One 2015; 10(10): e0139414.
[http://dx.doi.org/10.1371/journal.pone.0139414] [PMID: 26436944]

[73]   Sharma D, Lata M, Singh R, Deo N, Venkatesan K, Bisht D. Cytosolic proteome profiling of aminoglycosides resistant *Mycobacterium tuberculosis* clinical isolates using MALDI-TOF/MS. Front Microbiol 2016; 7: 1816.
[http://dx.doi.org/10.3389/fmicb.2016.01816] [PMID: 27895634]

[74]   Sharma D, Bisht D. tuberculosis hypothetical proteins and proteins of unknown function: hope for exploring novel resistance mechanisms as well as future target of drug resistance. Front Microbiol 2017; 8: 465.
[http://dx.doi.org/10.3389/fmicb.2017.00465] [PMID: 28377758]

[75]   Sharma D, Bisht D. Secretory proteome analysis of streptomycin-resistant *Mycobacterium tuberculosis* clinical isolates. SLAS DISCOVERY: Advancing Life Sciences R&D 2017; 22(10): 1229-38.

[76]   Sharma D, Bisht D. Role of bacterioferritin & ferritin in *M. Tuberculosis* pathogenesis and drug resistance: a future perspective by interactomic approach. Front Cell Infect Microbiol 2017; 7: 240.
[http://dx.doi.org/10.3389/fcimb.2017.00240] [PMID: 28642844]

CHAPTER 7

Overview and Challenges in Proteins Sample Preparation for 2-DGE in Bacterial Proteomics

Divakar Sharma[1,*], Juhi Sharma[2], Nirmala Deo[3] and **Deepa Bisht[3]**

[1] *CRF, Mass Spectrometry Laboratory, Kusuma School of Biological Sciences (KSBS), Indian Institute of Technology-Delhi, 110016, India*

[2] *Department of Botany and Microbiology, St. Aloysius College, Sadar Cantt, Jabalpur, 482002, India*

[3] *Department of Biochemistry, National JALMA Institute for Leprosy and other Mycobacterial Diseases, Tajganj, Agra, 282004, India*

Abstract: Sample preparation is the most crucial step in the proteome research of microbes. 2D gel electrophoresis (2-DGE) is a high throughput approach by which all proteins imprint on the gel in the form of spots or dots. Various protocols for the extraction of proteins exist in the literatures that are compatible with 2-DGE based proteome analysis from different microbes. Analysis of low abundance proteins, the solubility of proteins, and their resolution on the gel are some major issues observed with sample preparation and 2-DGE. To combat these issues, researchers have developed improved versions of the existing protocols using detergent/chaotropes for the enrichment of proteins during extraction along with the compatible chemical precipitation for better resolution of 2D gel pattern. In this chapter, we will discuss the overview, challenges encountered during the protein sample preparation, and their possible solutions in order to get a better 2-D gel of microbial proteins.

Keywords: 2-DGE, Detergents, Microbial proteins, Protein sample preparation.

INTRODUCTION

Protein sample preparation is one of the most crucial, yet problematic steps for high-quality resolution of proteins through two-dimensional gel electrophoresis (2-DGE) in microbial research. A lot of studies were conducted where high-throughput 2-DGE based approach was applied for the exploration of the various basics and clinical facts in different types of microbes [1 - 9]. Since the last twenty-five years, 2-DGE based approach has been used to analyze the whole microbial proteome as well as different fractions like a membrane, cytosolic, and

---

* **Corresponding author Divakar Sharma:** CRF, Mass Spectrometry Laboratory, Kusuma School of Biological Sciences (KSBS), Indian Institute of Technology-Delhi, 110016, India; Tel: +91-7906026680; E-mail: divakarsharma88@gmail.com

Divakar Sharma (Ed.)
All rights reserved-© 2020 Bentham Science Publishers

extracellular/secretory. Hence, 2-DGE has been considered the most accepted analytical method for resolving complex mixtures of proteins or expression proteomics [2 - 13]. Though various sample preparation protocols are available that are compatible with 2-DGE, researchers still keep on improving the strategies to get better 2D-gels [3, 14, 16]. In this chapter, we will discuss the overview and challenges faced during protein sample preparation and possible solutions to them.

Overview, Challenges and Probable Solutions

Sample preparation, which is the most crucial step for 2-DGE faces many problems and often encountered by the co-extraction of non-protein cellular components that can affect protein migration. While many methodologies have been described to alleviate these problems, defining optimal conditions for sample preparation from every new cell source is still somewhat a state-of-the-art process [14 - 16]. Proteins can be separated from non-protein contaminants by precipit-ation, or based on their characteristics, specific proteins can be enriched and then solubilized in sample buffer [3]. Because these techniques are often protein-specific or must be optimized for use, so a more universal approach for sample preparation is needed- one yielding a quality resolution of proteins in 2-D gels without laborious optimization and along with a higher tolerance for interfering non-protein components. Improved protein sample preparations have been used to reduce impurities as well to augment the low abundance proteins. Various types of selective protein extraction and protein precipitation with or without using detergents protocols were used to improve sample preparation for 2-DGE, which not only led to an increased number of spots but also increased resolution of the 2D gel patterns [3, 14 - 17].

Proteomics of lipophilic/membrane & associated proteins remains a major challenge in microbial research. Profiling of lipophilic proteins has been considered to be difficult within the bounds of conventional protocols for 2-DGE due to the limited solubility in aqueous buffer systems and relatively low abundance as compared to higher abundant cytoplasmic proteins. The nature of first dimension isoelectric focusing (IEF) requires solubilization of the proteins because they are subjected to an electric field in which they migrate to their isoelectric point. In addition to being highly hydrophobic, many integral membrane proteins tend to be very large and have transmembrane helices (TMH), which are major hurdles in the preparation of lipophilic/membrane protein samples for 2DGE [3, 16, 18].

Fractionation of cell extracts and subsequent analysis of these fractions remain a commonly used method for sample preparation. Several studies have reported the extraction of lipophilic/membrane and membrane-associated proteins using ultracentrifugation to obtain purified cell wall and cell membrane fractions for analysis by 2-DGE coupled with mass spectrometry [3, 18]. Common for these studies are the pre-isolation of the membrane and cell wall fractions from bacteria and the application of different washing techniques before protein extraction by detergents. Proteins could also be separated from non-protein contaminants by precipitation or based on their characteristics, specific proteins can be enriched and then solubilized in sample buffer [3, 18]. These major issues can be overcome by firstly, selective extraction of membrane proteins into a specific extraction buffer that contains IEF compatible detergent. Second, maintaining protein solubility throughout loading onto IPG strips and then subsequent first dimension IEF separation.

Our group has developed an efficient and rapid method to enrich the lipophilic proteins from *Mycobacterium tuberculosis* for 2-DGE [16]. The use of nonionic detergent (Triton X-100) in sonication buffer during extraction enriched the solubilization of the lipophilic proteins. Further, enriched whole cell lysate was treated with Triton X-114 and subjected to direct phase separation of lipophilic proteins, without the need for pre-isolation of membranes. The optimized extraction buffer increased the solubility of lipophilic proteins and improved resolution on 2-DGE as compared to standard extraction buffer and therefore considered as an efficient and rapid protocol for lipophilic proteins extraction and separation on 2-DGE.

Although highly efficient membrane protein extraction methods are routinely carried out with a detergent (SDS) for one-dimensional PAGE, due to the charged head group, SDS is incompatible with IEF experiments. To overcome this problem, SDS solubilized samples are subjected to precipitation by organic solvent or acid for removal of SDS and lipids contaminants [14]. Despite these harsh treatments followed by precipitation and delipidation by solvent, there was improved protein recovery without any modification.

Isoelectric focusing with immobilized pH gradients leads to severe quantitative loss of proteins in the resulting 2-D gel, although the resolution is usually high. Due to poor solubility, membrane proteins have not been properly resolved by high-resolution 2-DGE. To improve the solubility of proteins, various denaturing solution detergents (both nonionic and zwitterionic) and chaotropes (urea and thiourea) were used to get better results [16 - 19]. Various studies showed a significant improvement in the extraction of mycobacterial membrane proteins by enriched differential centrifugation and alkaline treatment of crude membranes with sodium carbonate and urea [20, 21].

Enrichment and Extraction of the Complete Proteome from Bacteria

Usually, the whole microbial proteins have been divided into three fractions; extracellular proteins or secretory proteins (proteins secreted by cell into the media), membrane and its associated proteins, and cytosolic proteins fractions. Membrane and its associated proteins and cytosolic fractions cumulatively make the whole cell lysate proteins. The bacterial cell wall is rigid, tough, and unable to break by normal cell lysis procedures like a human cell. Bacterial cell lysis was done by ultra-sonication and whole cell lysate was prepared [3 - 6, 8]. In brief, cell pellets were suspended in sonication buffer (1gm wet cell mass/ 5ml) containing PMSF (1mM), protease inhibitor cocktail (1 mM), and then broken by intermittent sonication for 2-20 minutes (for varieties of microbes) at 4°C in ice using a sonicator. Homogenate was cleared by centrifugation at 12,000 g for 15min.x 2 times at 4°C. The recovered supernatant was called whole cell lysate and used for proteomic investigations either directly or after performing precipitation protocols.

To enrich the membrane and its associated proteins in the cell lysate, cells were lysed by ultrasonication and cell lysate was prepared with slight modifications as described in our previously published manuscript [3]. In brief, cells were suspended in sonication buffer with 1% v/v Triton X–100 and then broken by intermittent sonication at 4°C for 2-20 minutes (for varieties of microbes). Homogenate was cleared by centrifuged at 12,000g for 20 minutes at 4°C. The resulting supernatant was further ultracentrifuged at 150,000 x g for 90 minutes and the pellet (membrane proteins) and supernatant (purely cytosolic proteins) were collected. Pellet (membrane proteins) was further washed and used for proteomic investigation directly by dissolving in rehydration buffer or further fractionated for intrinsic and extrinsic membrane proteins employing existing protocols. The fraction containing mainly cytosolic proteins was also used for proteomic investigation either directly or after protein precipitation.

It is also important to have an idea about the localization of proteins, as some of the proteins are secreted in culture medium known as secretory or culture filtrate proteins, so it should be checked in culture medium instead of cell lysate fraction. Recalling the culture filtrate or secretory or extracellular proteins from media is also a tedious task as these are usually secreted at very low concentrations. Apart from this, media contain salts and interfering molecules, which also pose a significant problem in good sample preparation for 2-DGE. For extraction of secretory proteins, previously published methods with slight modifications were followed [7, 15, 17]. In brief, bacterial cells grown in media up to their exponential phase were removed by centrifugation at 12000g for 20 minutes at 4°C. The supernatant (cell-free media) was first filtered by 0.45 μm and followed by 0.22-μm filters. The filtrate was further used for secretory protein extraction by the existing protein precipitation protocols.

Protein Precipitation Strategies

Based on the nature of proteins (whole cell lysate, pure cytosolic proteins, and secretory), various protein precipitation procedures are available in the literature [5 - 8, 15 - 21]. Types of bacteria also contribute to the selection of the precipitation method for the proteome analysis by 2-DGE. For whole cell lysate and cytosolic proteins usually, researchers used either SDS-TCA-acetone precipitation or cold-acetone precipitation procedures [5, 6, 8, 10, 19]. Both procedures are effectively used to analyze the 2D gel-based proteome in various bacteria, but due to the better resolution and more spots, the SDS-TCA-acetone precipitation protocol has been considered better than the cold-acetone precipitation. For culture filtrate or secretory proteins, usually, researchers used either the ammonium sulfate precipitation or SDS-TCA-acetone precipitation methods [7, 15, 17]. Both procedures are routinely used to extract secretory proteins but SDS-TCA-acetone precipitation has been considered the better over the ammonium sulfate precipitation because of salt minimization and improved sample preparation and good resolution of protein spots on 2D gels. All these strategies are briefly described below.

SDS-Tri Chloroacetic Acid (TCA)-Acetone Precipitation for Cell Lysate and Cytosolic Proteins

In the cell lysate and cytosolic proteins fraction (supernatant after ultracentrifugation), 10% SDS (w/v) was added to achieve a final concentration of 0.1% (w/v) and after boiling for 5 minutes, TCA was added to get 10% (w/v) final concentration [5, 6, 10, 19]. Mixtures were incubated at -20°C overnight for precipitation and precipitated protein pellets were collected by centrifugation (18,000 g, 15min, 4°C). Pellets were washed twice with one volume of 90% ice-cold acetone, followed by twice with 100% ice-cold acetone and pellet allowed to air dry and suspended in the appropriate volume of 2-DE rehydration buffer. Dissolved protein content was estimated by Bradford's method followed by 2-DE [23].

Acetone Precipitation for Cell Lysate and Cytosolic Proteins

The soluble protein fraction/whole cell lysate/cytosolic proteins fraction was precipitated overnight at -20°C by adding ice-cold acetone (100%) to the supernatant in excess or 1:4 ratio (1 volume sample: 4 volume ice-cold acetone) [8]. The precipitated proteins were collected by centrifugation (18,000 g, 15 min, 4°C). Pellets were washed twice with one volume of 100% ice-cold acetone and allowed to air dry and suspended in the appropriate volume of 2-DE rehydration buffer. The concentration of dissolved proteins was determined using the Bradford method followed by 2-DE [23].

Ammonium Sulfate Precipitation for Secretory/Culture Filtrate Proteins

After removal of the cells from media (4-week old culture) by centrifugation and filtration by 0.45 μm, followed by 0.22-μm filters; the filtrate was subjected to secretory proteins/culture filtrate proteins extraction by the ammonium sulfate precipitation protocols. In brief, culture filtrate or supernatant was concentrated by ammonium sulfate precipitation (80% saturation), and the salt was removed by dialysis [17, 22]. Pellets were washed twice with one volume of 100% ice-cold acetone and allowed to air dry and suspended in the appropriate volume of 2-DE rehydration buffer. The concentration of dissolved proteins was determined using the Bradford method followed by 2-DE [23].

SDS-Tri Chloroacetic Acid (TCA)-Acetone Precipitation for Secretory/Culture Filtrate Proteins

After removal of the cells from media (4-week old culture) by centrifugation and filtration as mentioned above, the filtrate was subjected to the SDS-TCA-acetone precipitation [7, 15, 17]. In brief, 10% SDS was added to the culture filtrate to achieve a final concentration of 0.1% (w/v) and after boiling for 5 minutes, and then TCA was added into the culture filtrate to get 10% (w/v) final concentration as in the tube. Mixtures were incubated at -20°C overnight for precipitation and precipitated protein pellets were collected by centrifugation (18,000 g, 30 minutes, 4°C). The minimum volume of HPLC grade water was added to the pellet and suspended. The whole suspension was washed with one volume of 90% ice-cold acetone, followed by twice with 100% ice-cold acetone, pellet allowed to air dry then suspended in the appropriate volume of 2-DE rehydration buffer. Protein content was estimated by Bradford's method, followed by 2-DE [23].

Isolation and Fractionation of Cell Membranes Proteins through Ultracentrifugation

The whole cell lysate was ultracentrifuged at 150,000 x g for 90 minutes and the pellet (membrane proteins) was collected [3, 18, 20, 21]. The recovered membrane pellet was resuspended in sonication buffer at a concentration of 10 mg/ml, and pre-condensed Triton-X114 was added, at a final concentration of 2%. The suspension was then stirred for 1 hr, at 4 °C to get the mixture in a single phase. Residual insoluble matter was removed by centrifugation at 150 000 × g and supernatant were allowed for phase separation in a water bath at 37 °C. The upper (aqueous) and lower (detergent) phases were collected by centrifugation

(1000 × g) and 'whole procedure was repeated thrice. Protein from the detergent phase was recovered by adding ice-cold acetone (100%) in excess or 1:5 ratios (1 volume sample: 5 volume ice-cold acetone). Pellet was resuspended in HPLC grade water, washed with 100% acetone, allowed to air dry, and suspended in the appropriate volume of 2-DE rehydration buffer. Dissolved protein content was estimated by Bradford's method, followed by 2-DE [23].

Effective Enrichment and Fractionation of Lipophilic Proteins without Ultracentrifugation

In 2016, our group published an efficient protocol for the effective enrichment of lipophilic proteins from mycobacteria by adding non-ionic detergent Triton X-100 at the time of extraction/sonication. After enrichment, whole cell lysate was subjected to direct phase separation using Triton X-114, without the need for pre-isolation of membranes [16]. In brief, for improvement of lipophilic protein solubility, Triton X-100 (1% v/v, Sigma) was added in the sonication buffer with bacterial cell and then broken by intermittent sonication for 5-15 min at 4°C. To remove the cell debris, homogenate was initially centrifuged at 2300 × g for 20 min at 4°C. Triton X-114 was added in the whole-cell lysate (final detergent concentration 2% v/v) and the suspension was stirred at 4°C for 20 minutes to obtain the protein extract in a single phase. Residual insoluble matter was removed by centrifugation at 15 700 × g for 10 min, and the solution was separated into two phases, an upper (aqueous) and a lower (detergent) phase after 10 minutes at 37°C. The detergent phase was collected and proteins were precipitated by 1:4 ratios of HPLC grade acetone. The protein pellet was washed and dissolved in the 2D rehydration buffer.

CONCLUSION AND FUTURE PROSPECTS

In this chapter, we have discussed that the protein sample preparation is a very crucial step in 2-DGE based proteome research. Usually, cytosolic, membrane, and its associated and secretory proteins constitute the whole cell proteins. Proteins of these fractions have their specific extraction and solubilization issues. Here, we have provided a glimpse of improved extraction, solubilization, and precipitation strategies for all the fractions, which provided a better outcome in the form of more numbers of protein spots and good separation or resolution of spots on 2-D gels. These improved strategies, coupled with 2-DGE could be used in future research to execute the expression-based problems.

CONSENT FOR PUBLICATION

Not applicable.

CONFLICT OF INTEREST

The authors declare no conflict of interest, financial or otherwise.

ACKNOWLEDGEMENTS

Declared none.

REFERENCES

[1]   Kumar B, Sharma D, Sharma P, Katoch VM, Venkatesan K, Bisht D. Proteomic analysis of *Mycobacterium tuberculosis* isolates resistant to kanamycin and amikacin. J Proteomics 2013; 94: 68-77.
[http://dx.doi.org/10.1016/j.jprot.2013.08.025] [PMID: 24036035]

[2]   Lata M, Sharma D, Kumar B, *et al.* Proteome analysis of ofloxacin and moxifloxacin induced *Mycobacterium tuberculosis* isolates by proteomic approach. Protein Pept Lett 2015; 22(4): 362-71.
[http://dx.doi.org/10.2174/0929866522666150209113708] [PMID: 25666036]

[3]   Sharma D, Kumar B, Lata M, *et al.* Comparative proteomic analysis of aminoglycosides resistant and susceptible *Mycobacterium tuberculosis* clinical isolates for exploring potential drug targets. PLoS One 2015; 10(10)e0139414
[http://dx.doi.org/10.1371/journal.pone.0139414] [PMID: 26436944]

[4]   Lata M, Sharma D, Deo N, Tiwari PK, Bisht D, Venkatesan K. Proteomic analysis of ofloxacin-mono resistant *Mycobacterium tuberculosis* isolates. J Proteomics 2015; 127(Pt A): 114-21.
[http://dx.doi.org/10.1016/j.jprot.2015.07.031] [PMID: 26238929]

[5]   Sharma D, Lata M, Singh R, Deo N, Venkatesan K, Bisht D. Cytosolic proteome profiling of aminoglycosides resistant *Mycobacterium tuberculosis* clinical isolates using MALDI-TOF/MS. Front Microbiol 2016; 7: 1816.
[http://dx.doi.org/10.3389/fmicb.2016.01816] [PMID: 27895634]

[6]   Qayyum S, Sharma D, Bisht D, Khan AU. Protein translation machinery holds a key for transition of planktonic cells to biofilm state in *Enterococcus faecalis*: A proteomic approach. Biochem Biophys Res Commun 2016; 474(4): 652-9.
[http://dx.doi.org/10.1016/j.bbrc.2016.04.145] [PMID: 27144316]

[7]   Sharma D, Bisht D. Secretory proteome analysis of streptomycin-resistant *Mycobacterium tuberculosis* clinical isolates. SLAS DISCOVERY: Advancing Life Sciences R&D 2017; 22(10): 1229-38.
[http://dx.doi.org/10.1177/2472555217698428]

[8]   Khan A, Sharma D, Faheem M, Bisht D, Khan AU. Proteomic analysis of a carbapenem-resistant Klebsiella pneumoniae strain in response to meropenem stress. J Glob Antimicrob Resist 2017; 8: 172-8.
[http://dx.doi.org/10.1016/j.jgar.2016.12.010] [PMID: 28219823]

[9]   Sharma D, Bisht D, Khan AU. Potential alternative strategy against drug resistant tuberculosis: a proteomics prospect. Proteomes 2018; 6(2): 26.
[http://dx.doi.org/10.3390/proteomes6020026] [PMID: 29843395]

[10]   Qayyum S, Sharma D, Bisht D, Khan AU. Identification of factors involved in *Enterococcus faecalis* biofilm under quercetin stress. Microb Pathog 2019; 126: 205-11.
[http://dx.doi.org/10.1016/j.micpath.2018.11.013] [PMID: 30423345]

[11]   Sharma D, Bisht D. tuberculosis hypothetical proteins and proteins of unknown function: hope for exploring novel resistance mechanisms as well as future target of drug resistance. Front Microbiol 2017; 8: 465.

[http://dx.doi.org/10.3389/fmicb.2017.00465] [PMID: 28377758]

[12]   Kumar G, Shankar H, Sharma D, *et al.* Proteomics of culture filtrate of prevalent *Mycobacterium tuberculosis* strains: 2D-PAGE map and MALDI-TOF/MS analysis. Slas Discovery: Advancing Life Sciences R&D 2017; 22(9): 1142-9.
[http://dx.doi.org/10.3389/fmicb.2017.00465] [PMID: 28377758]

[13]   Sharma D, Deo N, Bisht D. Proteomics and bioinformatics: a modern way to elucidate the resistome in *Mycobacterium tuberculosis*. J Proteomics Bioinform 2017; 10e33
[http://dx.doi.org/10.4172/jpb.1000e33]

[14]   Herbert BR, Molloy MP, Gooley AA, Walsh BJ, Bryson WG, Williams KL. Improved protein solubility in two-dimensional electrophoresis using tributyl phosphine as reducing agent. Electrophoresis 1998; 19(5): 845-51.
[http://dx.doi.org/10.1002/elps.1150190540] [PMID: 9629925]

[15]   Sharma D, Shankar H, Lata M, Joshi B, Venkatesan K, Bisht D. Culture filtrate proteome analysis of aminoglycoside resistant clinical isolates of *Mycobacterium tuberculosis*. BMC Infect Dis 2014; 14(S3): 60.
[http://dx.doi.org/10.1186/1471-2334-14-S3-P60]

[16]   Sharma D, Bisht D. An efficient and rapid method for enrichment of lipophilic proteins from *Mycobacterium tuberculosis* H37Rv for two-dimensional gel electrophoresis. Electrophoresis 2016; 37(9): 1187-90.
[http://dx.doi.org/10.1002/elps.201600025] [PMID: 26935602]

[17]   Kumar G, Shankar H, Bisht D, *et al.* A simple and rapid method of sample preparation from culture filtrate of *M. Tuberculosis* for two-dimensional gel electrophoresis. Braz J Microbiol 2010; 41(2): 295-9.
[http://dx.doi.org/10.1590/S1517-83822010000200005] [PMID: 24031494]

[18]   Sinha S, Arora S, Kosalai K, Namane A, Pym AS, Cole ST. Proteome analysis of the plasma membrane of *Mycobacterium tuberculosis*. Comp Funct Genomics 2002; 3(6): 470-83.
[http://dx.doi.org/10.1002/cfg.211] [PMID: 18629250]

[19]   Bisht D, Singhal N, Sharma P, Venkatesan K. An improved sample preparation method for analyzing mycobacterial proteins in two-dimensional gels. Biochemistry (Mosc) 2007; 72(6): 672-4.
[http://dx.doi.org/10.1134/S0006297907060119] [PMID: 17630913]

[20]   Gu S, Chen J, Dobos KM, Bradbury EM, Belisle JT, Chen X. Comprehensive proteomic profiling of the membrane constituents of a *Mycobacterium tuberculosis* strain. Mol Cell Proteomics 2003; 2(12): 1284-96.
[http://dx.doi.org/10.1074/mcp.M300060-MCP200] [PMID: 14532352]

[21]   Xiong Y, Chalmers MJ, Gao FP, Cross TA, Marshall AG. Identification of *Mycobacterium tuberculosis* H37Rv integral membrane proteins by one-dimensional gel electrophoresis and liquid chromatography electrospray ionization tandem mass spectrometry. J Proteome Res 2005; 4(3): 855-61.
[http://dx.doi.org/10.1021/pr0500049] [PMID: 15952732]

[22]   Nagai S, Wiker HG, Harboe M, Kinomoto M. Isolation and partial characterization of major protein antigens in the culture fluid of *Mycobacterium tuberculosis*. Infect Immun 1991; 59(1): 372-82.
[http://dx.doi.org/10.1128/IAI.59.1.372-382.1991] [PMID: 1898899]

[23]   Bradford MM. A rapid and sensitive method for the quantitation of microgram quantities of protein utilizing the principle of protein-dye binding. Anal Biochem 1976; 72: 248-54.
[http://dx.doi.org/10.1016/0003-2697(76)90527-3] [PMID: 942051]

CHAPTER 8

Proteomics Based Biomarker Development Against Infections: An Overview

M. Madhan Kumar[1,*], Vivek Kumar Gupta[1] and **Divakar Sharma[2]**

[1] *ICMR-National JALMA Institute for Leprosy and other Mycobacterial Diseases, Tajganj, Agra-282004, India*

[2] *CRF, Kusuma School of Biological Sciences, Indian Institute of Technology, Delhi*

Abstract: Pathogens like viruses, bacteria, protozoa, fungi and helminths have caused and are causing diseases as well as the death of mankind since time immemorial. A proper understanding of the infection mechanisms will aid in designing efficient vaccines, diagnostics and drugs for efficient disease management. When compared to genes, proteins are involved in effector functions of the cell (of the pathogen as well as the host) and play a pivotal role in pathogen entry into host cells, its transmission and disease pathogenesis. Hence, the study of proteins will decipher the intricate communication networks involved in disease pathogenesis and will also elucidate their use as biomarkers in diagnostic and therapeutic scenarios. Thus, this chapter will highlight the proteomic and mass spectrometry-based approaches in comprehending the pathogen interactions with the host at the molecular level as well as biomarker development for use in diagnostics, prognostics and disease management.

Keywords: Biomarkers, Diagnostics, Mass spectrometry, Infectious diseases, Proteomics, Therapeutics.

INTRODUCTION

Proteins and Their Role in Infectious Diseases

Life is the Mode of Action of Proteins - Friedrich Engels [1]

Proteins form an integral part of life, whether in nutrition or in pathogen defence/pathogen attack. Proteins from the host in the form of cytokines, immunoglobulins, complement or other immune components (immune peptides like defensins, cathelicidins *etc.*), mount an attack against the pathogen and the pathogen, in turn, produces endotoxins or exotoxins to cause damage to the host. Moreover, pathogens also utilize the host cell components (like lipids, proteins,

---

[*] **Corresponding author M. Madhan Kumar:** ICMR-National JALMA Institute for Leprosy and other Mycobacterial Diseases, Tajganj, Agra-282004, India; Tel: +91-562-2331756; E-mail: madhanbiophile@gmail.com

Divakar Sharma (Ed.)
All rights reserved-© 2020 Bentham Science Publishers

carbohydrates) for their survival and propagation. But, for successful survival, the pathogen must usurp the protein machinery of the host, or suppress the functioning of host proteins [2]. It is because, in the host, protein interactions play a pivotal role in cellular processes like transcription, translation, cell communication, cell adhesion, protein synthesis, cell signaling, *etc.* A disruption of this protein-protein interaction (PPI) network by the invading pathogen/establishment of new networks with host proteins by the pathogens play a major role in the progression of pathogenesis and the establishment of disease. Thus, a detailed study of PPI will unveil the pathological mechanisms underlying infection and will also aid in devising strategies for targeting the pathogen, help in disease diagnosis as well as prevention [3]. The protein-protein interactions can be direct (wherein the protein directly interacts with other protein) or can be indirect (the protein interacts with the other protein *via* intermediate molecules). A detailed study of this cross-talk between proteins (between host proteins or between host-pathogen proteins) in a cell has been termed as 'interactome'. The PPI networks have been studied in *Yersinia pestis, Bacillus anthracis, Franciscella tularensis* in order to define the proteins involved in pathogenesis and the associated mechanisms [4].

One of the challenges in interactome studies is the dynamic change in protein profile at different stages of the disease, which reflects the immune response of the host to the pathogen and also the expression of different virulent proteins by the pathogen during the course of disease [5]. Thus, the study of such interactions also takes into consideration the time as well as space for getting thorough knowledge about the interactomes (spatio-temporal analysis). Apart from the perspective of studying proteins in host-pathogen interactions, the study of proteins of the pathogens will help in discriminating between species and strains.

Apart from this aspect, functions and protein interactions are regulated by post-translational modification (PTM) of proteins. Such regulation by PTM on the host or pathogen proteins plays an important role in the infection progression and outcome. Regulation of host and pathogen PTM signalling deserves importance during infection as it impacts the changes in protein interactions, subcellular localization and, ultimately, the replication of pathogens. Among the PTMs, phosphorylation plays an important role as the pathogens are known to encode kinases, which can manipulate the host kinase signalling pathways. Thus, if proteomic technologies are applied efficiently in studying PTMs, they will help in unravelling pathogen-specific PTM signalling cascades, which can be targeted in therapeutics for a favourable outcome.

Although proteomics *per se* has contributed a lot to understanding proteins, integrating it with other molecular biology and biochemical methods has yielded more information about infections. A combination of different 'omics' technologies like genomics, transcriptomics, complements the information obtained from proteomics and has made the understanding of infection a better way. Thus, multiple 'omics' technology approaches have given an enhanced insight into the process of pathogenesis in the case of infectious diseases.

Techniques Used for Studying Protein-protein Interactions (PPI)

The most common technique used for studying PPI for studying host-pathogen interactions is immunoaffinity purification coupled with mass spectrometry (IP-MS). This method has helped in identifying proteins involved in the interaction between pathogens and the host. It has also aided in elucidating cellular defense mechanisms of the host [6]. The method involves isolating the proteins by using antibodies raised against them. The antibodies are bound to sepharose or agarose beads in which the cell suspension containing the protein of interest is passed through, and the unbound proteins are washed off. Once the specific protein gets bound to the antibody, it is eluted by using competing epitope peptides for the antibody or by varying the pH or by the use of salts to dissociate the antigen-antibody complex. The dissociated protein is then subjected to mass spectrometry to identify the proteins. Thus, this technique can be useful for enriching proteins of low abundance, identifying unknown proteins in the complex. In Immunoprecipitation of protein complexes, the antibody (primary) is directed towards one of the proteins in the complex. Once the antibody is bound to the protein of interest, sepharose beads containing secondary antibodies are added to this mixture, wherein the secondary body attached to the beads bind to the primary antibodies bound to the protein complexes. This is centrifuged (washing) to precipitate the bead containing protein complexes from other proteins. The protein complexes are then eluted using the methods as said above, and then the proteins are subjected to mass spectrometry for identification to know about the interacting proteins in the complexes (Fig. **1**).

In certain circumstances, the proteins of interest are tagged with epitopes, and then antibodies against the epitopes are used for isolating the proteins (wherein no antibodies can be raised against a particular protein, epitope tagging is employed). The epitope tagging method is useful for studying proteins, which are very low or for studying novel proteins. After this process, the purified proteins are identified by the proteomic approach wherein the proteins are cleaved into peptides (by trypsin) and identified by mass spectrometric approaches.

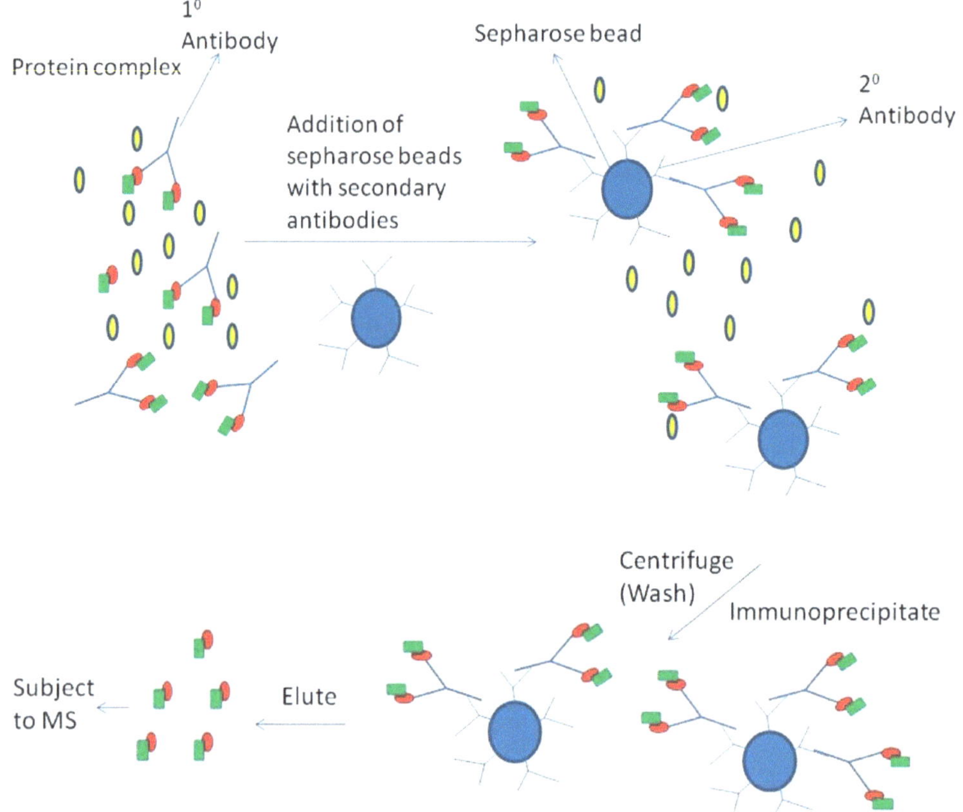

Fig (1). Immunoprecipitation of Protein complexes.

To complement IP-MS approaches, high throughput screening techniques like Nucleic Acid Programmable Protein Array (NAPPA) have shown promising leads for pathogen interactions. The technique employs the plasmid DNA of the protein instead of proteins used in microarrays. This technology uses a cell-free expression system, *i.e.*, cell extracts are used, which provide raw materials for transcription, translation machinery (as protein synthesis can happen without the need for an intact cell). The protein-encoding plasmid DNA is spotted in a protein capture surface and translated using a cell-free expression system prior to sample application. This technology obviates the need to purify a protein, instead one needs to produce the DNA clones which encode the proteins (Fig. **2**).

The NAPPA technology has been used to express viral proteins from different virus species for detecting antibodies in human serum specific to viral epitopes [7]. It has also been used to study the direct interactions of *Legionella pneumophila* proteins – LidA and SidM interaction with 10,000 human proteins (imprinted on NAPPA) [8]. Apart from infectious diseases, the NAPPA has also found use in cancer, autoimmune diseases.

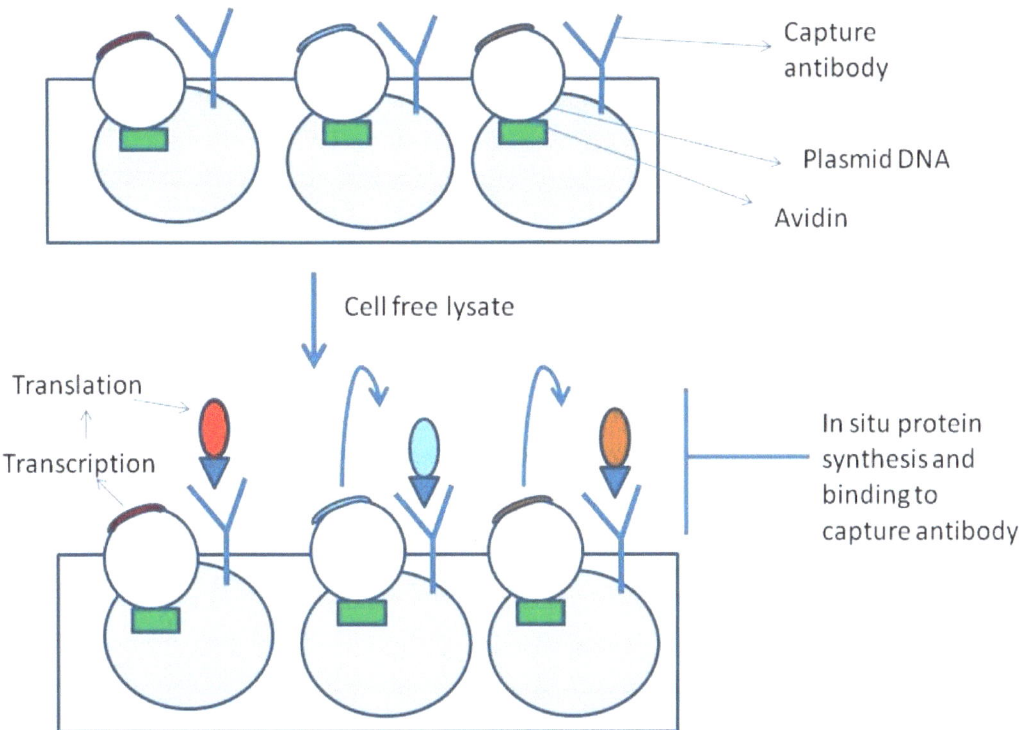

Fig. (2). Nucleic acid programmable protein array (NAPPA).

Another important method used for studying PPI is Yeast two-hybrid (Y2H) system (Fig. **3**). It exploits the principle that eukaryotic transcription factors contain DNA binding domain (DBD) and a transcription activation domain (TAD). The interaction between a bait protein-containing DBD and a target protein of interest fused to TAD leads to transcription of a downstream reporter gene. The methodology involves the construction of 2 plasmids: one containing the bait protein fused to DBD and the other containing the target protein fused to TAD. Both the plasmids are co-expressed in the same cell. If there will be a protein interaction between the bait with the target, transcription will be initiated, and a downstream gene will be expressed. But, Y2H suffers from many demerits like: a) ability to detect binary interactions only b) Bait and target fusion proteins are targeted to the nucleus; hence proteins containing signals for targeting to other compartments cannot be explored c) If the protein interaction happens after a post-translation modification, such interactions cannot be detected d) Folding of proteins in yeast will be different from that in humans, hence due to improper folding, the proteins may not bind to the target leading to no interactions. Thus, the Y2H system can give false positives and false negatives to a great extent.

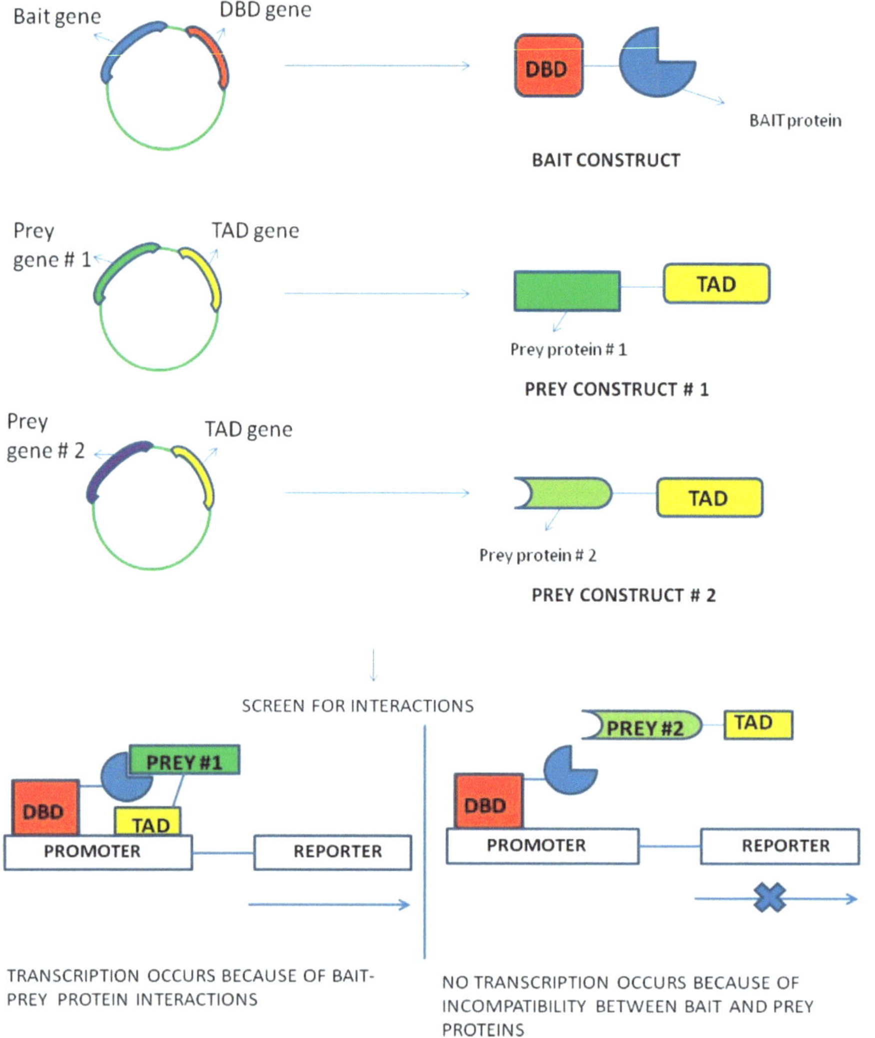

Fig. (3). Yeast 2 Hybrid system for studying protein interactions. DBD – DNA Binding Domain; TAD – Transcription Activation Domain.

Protein microarrays have also been found to be invaluable in PPI detection, wherein a fluorescent signal is used for the detection of interactions [9]. Although this technology can detect a large array of protein interactions using a small amount of sample, it suffers from drawbacks like inability to retain post-translational events, protein immobilization, and transient protein interactions being lost.

For quantification of protein levels, IP-MS technology has relied on labelling approaches like Stable isotope labelling of amino acids in cell culture (SILAC) and isobaric tags or label-free approaches like spectral counting [10]. Many studies have combined labelling approaches with label-free IP-MS for having a greater effect.

In Silico Methods for Studying Protein-protein Interactions

Apart from the above mentioned *in vitro* methods, *in silico* methods have also been used for studying protein-protein interactions. The computational predictions are made by using the available structural information [11] or by the use of genomic data [12] or by algorithms [13]. Those PPIs which have been experimentally detected by labs are also found in databases like biomolecular interaction network database (BIND), Munich information centre for protein sequences (MIPS), database of interacting proteins (DIP), and Search Tool for the Retrieval of Interacting Genes/Proteins (STRING) [11, 14 - 17]. Many are user-friendly but suffer from inherent demerits of false positives and negatives from the information deduced from datasets. Thus, they can be used for validating results obtained from experimental studies. After the use of *in silico* methods, the next step is to experimentally detect it. This is done by the use of high throughput techniques like Y2H/IP-MS/protein microarrays *etc*. Apart from these methods, there are also other methods like electrophoretic mobility shift assay (EMSA) [18], phage display [19] and microscale thermophoresis [20], which find use in studying PPI (which have not been discussed in this chapter).

A Note on Biomarkers and Their Role in Infectious Diseases

Infectious diseases, if detected early, can play an important role in the management of the disease and subsequently can lead to reduced morbidity and mortality associated with it. In the case of detection of bacterial resistance, this deserves much importance as early diagnosis of resistance based on biomarkers will help in devising strategies for altering therapy on an individual basis for tackling resistance problems. Biomarkers are useful indicators for biological processes, knowing pathogenic processes and pharmacological processes and aid in therapeutic interventions. The biomarkers are categorized into diagnostic, prognostic and predictive types. Diagnostic biomarkers play a role in detecting a disease desirably at an early stage. Prognostic biomarkers help in monitoring the development and extent of disease. Predictive biomarkers play a role in monitoring clinical response as well as toxicity associated with a treatment [21]. In infectious diseases, biomarkers can be derived from the pathogen or the host.

Challenges in Studying Protein-protein Interactions

One of the main challenges in studying protein-protein interactions using the above-said methodologies is the non-specific binding, which should be differentiated from specific binding. Moreover, transient protein interactions and the interactions which depend on post-translational modifications are difficult to detect because of the inefficiency of the many available techniques to do this. In some cases, the protein does not interact directly with another protein but rather does it by means of other binding partners in the complex. Knowledge of this may help in achieving success in detecting PPI.

Protein Markers in Bacterial and Viral Infections – Pathogen and Host Perspectives

Viruses have either single-stranded/double-stranded RNA/DNA as their genome (but never RNA and DNA together). The study of viral proteins is relatively easy because of their relatively simple genome and production of few proteins. For efficient spread, viruses depend on regulated interactions with the proteins of the host (both in space and time). The host proteins targeted by the viruses have helped in improving our understanding of the function of viral proteins. In order to study viral proteins, the proteins must be present in abundance for easy isolation and for finding out the proteins associated with the viral proteins (host/viral).

Viruses, after infection, hijack the host cell machinery for survival and replication. This hijacking leads to changes in host gene expression. These changes can be monitored and investigated for identifying potential biomarkers. For identifying biomarkers in viral infections, healthy and diseased specimens are collected, protein expression between the two types of samples are compared. The differentially expressed proteins are validated for discovering new biomarkers. For analyzing purified virions, matrix-assisted laser desorption ionization (MALDI) coupled with time of flight (TOF) mass spectrometry (MS) and liquid chromatography (LC) linked with tandem MS (MS/MS) has been widely used, which has added more to the knowledge of components of viruses. Clinically important viral infections which have been explored using mass spectrometry-based proteomics for the search of biomarkers are as follows:

Influenza

Influenza is a negative-sense RNA virus that is of three different types: A, B and C. Among the three, Influenza A poses a great threat to humans because of its mutating ability, and earlier there have been outbreaks of the subtypes like H5N1 [22], H2N2 [23], H1N1 [24]. A new influenza strain H1N1 emerged with genes included from human, avian and swine species by the process of antigenic shift.

Although strategies like the use of small molecules targeting the virus or vaccines have been used, they are found to be of little use due to the mutations which happen in the viral genome and the subsequent resistance developed against these arsenals. Thus, a detailed understanding of the host factors which permit the replication of the virus may provide a better strategy to control this infection.

Among the proteins of influenza A virus, hemagluttinin is an important glycoprotein because of its ability to bind to sialic acid receptors on host cells and helps it to attach to host cells. Apart from this protein, there are 10 other protein components encoded by influenza viruses like neuraminidase, M1, M2, NP, PA, PB1, PB1-F2, PB-2, NS1 and NS2 (or nuclear export protein).

When Influenza A and human PPI networks were explored, viral proteins were interconnected to a greater extent and were also found to be interacting with a large number of human proteins. This PPI network map has helped in identifying the numerous mechanisms used by the viruses for manipulating host cells. This map has also helped in knowing how viral proteins alter interferon production by the host [25].

A study by Tisconcik-Go *et al.* (2016) investigated host immune response after infection with different influenza strains. Different –omics datasets were used (proteomics, metabolomics, lipidomics, transcriptomics), and an integrative network was generated for identifying relationships between different omic datasets. This study has revealed there were multi-omic associations which were strain-specific [26].

Another study used a novel proteomics technology called SOMAScan, and about 1030 proteins from nasal mucosa lavages of patients were quantified using this technique. It was observed that out of 160 differentially expressed proteins between infected and uninfected individuals, many proteins were involved in the recruitment of immune cells (CXCL11, CXCL10, IL-16) [27].

Hepatitis C

Hepatitis C virus (HCV) is known to cause diseases like cirrhosis, hepatitis, hepatocellular carcinoma, and because of the morbidity and mortality it causes, it deserves much importance [28]. It is a single-stranded positive-sense RNA virus. Still, its life cycle is not known clearly; no vaccine is available against this virus. The virus is known to encode 9 proteins *viz.*, four structural proteins (core protein, E1, E2, P7) and five non-structural proteins. Many proteomic studies focus on the role of these 9 proteins with the host cells for deciphering the pathogenesis of this virus. Most of the proteomic studies have been done by infecting cell lines and thereafter, the expression of proteins by the cells. Tsutsumi *et al.*, (2009) infected

HepG2 cells with HCV and have found that 16 proteins were differentially expressed. Among these, 13 proteins increased and 3 proteins decreased. Among the host proteins, Prohibitin (a mitochondrial chaperone) was found to be upregulated after HCV protein interactions, and this was found to inhibit mitochondrial cytochrome C oxidase (COX), which lead to increased expression of the mitochondrial respiratory chain. These observations throw light on the knowledge regarding the generation of reactive oxygen species generation and the associated liver damage [29].

In a study by Kou *et al.*, (2006), host proteins interacting with non-structural protein 4 A (NS4A) or non-structural protein 4 B (NS4B) of HCV were investigated in Huh7 cells. It was observed that human translation elongation factor 1 alpha 1 (eEF1A) that signals aminoacyl tRNA to the ribosome was found to bind NS4A and not NS4B. Thus, this interaction diminished the effect on translation activities. The technique employed by the group for this study was the GST-pulldown assay coupled with LC-MS/MS [30].

Dengue

Dengue is an arthropod-borne disease (mosquitoes) caused by single-stranded RNA positive-strand viruses. The disease causes symptoms like headache, fever, joint pain and sometimes hemorrhagic fever (which is lethal). Till now, no treatment or vaccine is available for treating dengue. To find out target proteins which can be used as biomarkers, several proteomic studies have been done for characterizing the host and dengue virus-associated proteins during infection. In a study that investigated the plasma of severe dengue patients compared to healthy controls, it was observed that 73 proteins were differentially expressed, of which 37 were identified by mass spectrometry. Many pro-inflammatory proteins like ApoJ, Complement C3 and fibrinogen γ chain were all upregulated. About 5 anti-inflammatory proteins showed a decreased expression, whereas three showed an increased expression. The study was done using a difference in gel electrophoresis method (DIGE) coupled with MS [31]. In a study by the research group of Higa (2008), it was observed that in control and infected liver cells, 20 proteins were identified by gel electrophoresis and MS approaches. Out of these, 2 proteins were present in infected cells, namely thioredoxin and tissue inhibitor of metalloproteinase -1 (TIMP-1), the expression of which decreases in infected cells. This decrease might probably lead to increased endothelial permeability and plasma leakage because of TIMP-1 exhibiting anti-metalloproteinase activity [32].

Human Immunodeficiency Virus -1 (HIV-1)

HIV-1 is a retrovirus (in which RNA is the genetic material which gets reverse transcribed to DNA and gets integrated into host DNA from which viral proteins are produced which get assembled as virions) which targets CD4 immune cells, replicates in the cells by using the molecular machinery of CD4 cells/macrophages and finally destroys them. This destruction eventually leads to a reduction of counts of CD4 immune cells and subsequently, the individual develops immune deficiency. This ultimately leads to the development of acquired immune deficiency syndrome (AIDS), which makes the person susceptible to infections and finally leads to the death of the individual. Hence, this disease has deserved special attention, and much research has been done to understand more about the protein interactions and hence pathogenesis of the disease. Moreover, HIV-1 penetrates the brain and causes cognitive impairment and also death due to dementia. Thus, proteomic studies have also focused on differences in protein expression in individuals with and without dementia. In one such study, two proteins, namely gelsolin and prealbumin, were identified in individuals with and without cognitive impairment [33]. In another study, individuals with cognitive impairment showed an upregulation of gelsolin protein along with other proteins like clusterin, complement C3, VDBP, Cystatin C, Procollagen C-endopeptidase enhancer 1. This study used DIGE and LC-MS/MS for protein identification [34].

Severe Acute Respiratory Syndrome Coronavirus 2 (SARS-CoV-2)

The recent pandemic coronavirus disease 19 (COVID-19), which has wreaked havoc in the lives of humans, is caused by the virus severe acute respiratory syndrome coronavirus 2 (SARS-CoV-2). The morbidity and mortality associated with the disease, call for efficient biomarkers, which can be used in the diagnosis and those which can be targeted in therapeutics. In diagnostics, currently, real-time PCR (or quantitative PCR (qPCR) based tests (targeting different genes of the virus) are used worldwide for diagnosis. But, it suffers from sensitivity and specificity issues. The presence of less viral load, changes in the viral strain sequences, inability to detect between live and dead viruses (as it is nucleic acid-based) are special challenges which decrease the sensitivity of the qPCR based assays. Apart from these, the high costs of nucleic acid-based tests preclude their common use. The use of biomarkers, which can help in early diagnosis of the disease, predict those individuals who will progress to severe disease will be of utmost use to the medical community to start timely therapy to reduce the disease severity and mortality.

Biomarkers may also help in identifying drug targets as well as in the detection of the drug response (whether efficient or inefficient). This deserves attention in COVID-19 because of the unavailability of efficient drugs for combating this disease. Shen *et al.*, (2020) characterized the proteomic and metabolomic profiles in COVID-19 patient's sera. The study showed that 93 proteins exhibited differential expression in COVID-19 patients' sera (compared to healthy controls). The study group also observed that 204 metabolites in COVID-19 patient sera showed a correlation with disease severity. Characteristic protein and metabolite changes were observed in sera of COVID-19 patients, which the authors suggest might be useful for the selection of biomarkers for evaluating disease severity. They also suggest that clues for therapeutic targets may also be obtained from the studies on blood molecular changes observed between the groups. For proteomic investigation of sera, the authors employed reverse phase (RP)/ultra performance liquid chromatography (UPLC) mass spectrometry (MS) techniques coupled with electrospray ionization (ESI) [35].

Another study utilized the gargled solution from COVID-19 patients and identified SARS-CoV2 nucleoproteins by using a MS-based method. The method involved precipitation of proteins using acetone followed by tryptic digestion and MS analysis. Such MS based methods for detecting SARS-CoV2 peptides might be helpful in the future for early, reliable, diagnosis of COVID-19 [36]. Another interesting study has made a human cell culture model and infected with SARS-CoV2 and studied the translatome (entire set of mRNA fragments that are translated in a cell) and proteomics at different times after the infection. This study showed that the virus reshaped the metabolism, molecular processes like splicing, translation as well as protein homeostasis. The investigators of the study used small molecule inhibitors and found that targeting the pathways prevented the replication of SARS-CoV-2 in cells. The authors suggest that such dissection of molecular mechanisms will help in providing insights for devising therapies against this disease [37].

Thus, these studies point out that proteomic investigation will pave the way for designing diagnostics or therapies by targeting molecular pathways. Although not much has been done with regard to proteomic studies on SARS-CoV-2, detailed proteomic studies may provide a better understanding of the COVID-19 pathogenesis and affordable solutions in diagnostics and therapeutics for controlling this dreadful pandemic.

Search for Protein Markers in Bacterial Infections – Host and Pathogen Perspectives

Bacteria are also known to cause many infections in humans with a high index of morbidity and mortality. Bacterial resistance is another persistent problem, which has hampered the killing of bacteria by antibiotics. Thus a proper understanding of bacterial physiology will help in the rational design of drugs, vaccines and diagnosis against bacterial pathogens. Proteomics has played an important role in deciphering the proteins of bacteria and host involved in pathogenesis and their interactions for a better understanding of these pathogens and their impact on human health. Thus proteomics has helped a lot in studying post-translational modifications in bacteria, antimicrobial resistance, pathogen-host cell interactions as well as in the discovery of biomarkers.

Use of Bacterial Markers for Typing Bacteria

Bacterial markers will be of use if they can differentiate pathogenic species from non-lethal non-pathogenic species. Along these lines, a study has shown that ribosomal proteins (L15, L17, L21, L25, S7, S8, SodA, PPIAse C) of *Salmonella* can be used as markers for typing *S. enterica* subspecies enterica [38]. Another group has used glycolipids as markers for the identification of pathogens of ESKAPE (*Enterococcus faecium, Staphylococcus aureus, Klebsiella pneumoniae, Acinetobacter baumannii, Pseudomonas aeruginosa, Enterobacter spp.*) group [39]. With regard to the identification of bacteria, it has been observed that MALDI-TOF based mass spectrometry has the ability to identify gram-negative bacteria at the species level in an efficient way when compared to gram-positive bacteria. Many MS-based instruments are commercially available for the rapid identification of microorganisms. These instruments are of high-speed and are highly reliable for identifying and classifying bacteria, yeasts and fungi. These methods are successful because of the availability of databases, which are updated constantly after the discovery of new species. Thus the use of these technologies has helped in early diagnosis and treatment and hence has reduced the costs related to hospitalization [40, 41].

Determining Antibiotic Resistance

Determining antibiotic resistance in patients is a very important aspect of treatment. A clear knowledge about the antimicrobial susceptibility and resistance profiles will be useful in deciding the course of treatment for patients. The conventional antibiotic susceptibility testing methods include disk diffusion, agar and broth dilution methods. These tests determine the effect of drugs on the growth of microorganisms. The disadvantage with these tests is that they are time-consuming and this delays the treatment of the patient, which can lead to increased morbidity and mortality. Some of the ways by which bacteria become resistant are by the production of enzymes that can destroy the antibiotic by hydrolysis or chemical modification. The example for the former is β-lactamases produced by β-lactam resistant bacteria, which can destroy the β-lactam ring of the antibiotic and make it inactive [42]. Those bacteria resistant to chloramphenicol, produce acetyltransferases which acetylate the antibiotic and make it ineffective [43]. One another mechanism used by antibiotic-resistant bacteria is decreasing the permeability of antibiotics and preventing its entry into intracellular/periplasmic target molecules or by upregulating efflux pumps, a mechanism for exporting the antibiotic out of the cell [44].

To analyze growth rate and for building resistance profile, semi-quantitative MALDI-TOF MS has been successfully used for Mycobacteria, *Staphylococcus aureus*, gram-negative bacteria [45]. Ho *et al.*, (2017) have used MALDI-TOF along with vector machine models to generate an algorithm for identifying *Bacterioides fragilis* as well as for differentiating between CfiA (carbapenemase gene responsible for cleaving carbapenem group of antibiotics) positive and negative groups [46].

The bacterium *Pseudomonas aeruginosa* is known to cause chronic and acute infections in humans and multidrug resistance is a very big therapeutic challenge in tackling this pathogen [47, 48]. A selected reaction monitoring (SRM) based method has been employed to quantify proteins involved in the antibiotic resistance of *P. aeruginosa* like efflux pumps, OprD porins, AmpC cephalo-sporinase [49]. This research group has also studied two proteins (PBP2a, PBP2c) related to resistance in methicillin-resistant *Staphylococcus aureus* for β-lactam resistance using SRM method.

Apart from determining antibiotic resistance, proteomics has also been employed to study the evolution of resistance in bacteria. In this context, proteins secreted by lineages of mycobacteria have been studied in the presence of the antibiotic Rifampicin. For quantifying the protein levels between the lineages, data-dependent acquisition (DDA) followed by parallel reaction monitoring (PRM) was used. Interestingly, DosR (dormancy regulon) related proteins were present in abundance in the Beijing lineage of *Mycobacterium tuberculosis* when compared to other strains before Rifampicin treatment. This showed that DosR proteins in Beijing strains helped in persistence, thus providing it a better advantage in survival when compared to other strains. If the bacterium undergoes any assault, due to drugs or immune response, the chances of the strain undergoing dormancy will be more, thus increasing its chances of survival [50].

Proteomics in Biomarker Discovery and Diagnostics

Prompt diagnosis of bacterial infections is the key to start treatment to avoid other complications due to delay. New biomarker development will aid in the current diagnostic and therapeutic scenario and will help in betterment. Proteomics has helped in identifying potential biomarkers from different samples.

In a study on mycobacterial proteins, multiple reaction monitoring (MRM) was used for exosomes isolated from human serum samples of culture-confirmed TB patients, which identified 76 peptides (which represented 33 proteins of MTB). Of these 33 proteins, 20 proteins were found in exosomes isolated from the serum of TB patients [51]. The research group of Kruh-Garcia has shown by MRM that the MTB protein Ag85B showed variation in expression as well as secretion across different lineages and within sub-clades. This highlights the importance of MRM in distinguishing the variation in homologous proteins and also the use of this technology for aiding in the selection of biomarkers [52].

For identification of bacteria, targeted proteomics has been used along with other detection methods. The study by Larson *et al.*, (2013) has focussed on PCR as well as MRM methods for differentiating between highly virulent *Yersinia pestis* and the less virulent *Yersinia pseudotuberculosis*. The authors identified a gene *YPO1670*, which was unique to *Y.pestis*, which was confirmed using the PCR method. The protein encoded by the gene YPO1670 was monitored by the MRM method on *Y. pestis* samples. The study concludes that YPO1670 is a reliable marker for the identification of *Y. pestis* [53].

Biomarkers in Diagnosis, Treatment and Prevention of Infectious Diseases

As already described, biomarker development involves many phases starting from the design of a research experiment to discovery, verification and validation. The disease-specific biomarker studies have mostly focused on the high throughput mass spectrometry techniques for analysis. Separation methods used for the samples include liquid chromatography, gel electrophoresis (1 dimensional as well as 2 dimensional) coupled with mass spectrometry, MALDI-TOF MS, surface enhanced laser desorption ionization-TOF/MS (SELDI-TOF/MS), fourier transform ion cyclotron resonance/MS (FTICR-MS). In the case of biomarker identification, MRM and SRM have been used in combination with MS in the process of biomarker identification. For validation, MRM/SRM, enzyme-linked immunosorbent assay (ELISA), western blot, immunohistochemistry have been used. Doing global proteomic profiling of the proteins present in body fluids can reveal valuable biomarkers that might be of use in diagnosis, treatment and prevention of infectious diseases.

THE WAY AHEAD – FUTURE PERSPECTIVES

In infectious diseases, pathogens have learned how to evade the immune system and cause disease. Thus immunotherapy, which targets the immune system, plays a role in manipulating it to make it robust to fight against the invading pathogen. Mass spectrometry aids in identifying disease-related peptides/metabolites, which can be later investigated for their use as targets in immunotherapy. The MS technology has played a vital role in identifying biomarkers for diseases, aided in designing T cell therapies for infections and even designing vaccines. But the caveat is MS-based identification, which is the individual-specific immune responses that need to be addressed. Thus the use of MS towards the identification of universal antigens/biomarkers will be of more relevance rather than the personalized methods. For efficient combating of the disease, effective immunotherapy needs to be formulated and for achieving this, proper identification and understanding of the disease is a pre-requisite, which MS-based technologies may fulfill. Many challenges are faced with respect to biomarker discovery using MS technology like a low abundance of proteins, complexity in the body fluids, *etc.* To address these issues in an efficient way, proteomics alone may not suffice. The help of other –omics technologies like metabolomics, peptidomics, genomics aided by bioinformatic methods may help in bridging this gap and afford effective solutions. Thus, the combination of other –omics approaches will give deeper insights into the proteome landscape and help devise efficient strategies in our combat against infectious diseases.

CONSENT FOR PUBLICATION

Not applicable.

CONFLICT OF INTEREST

The authors declare no conflict of interest, financial or otherwise.

ACKNOWLEDGEMENTS

The authors would like to acknowledge Dr. Shripad A Patil, Director, National JALMA Institute for Leprosy and other Mycobacterial diseases for his support and encouragement in the completion of this chapter.

REFERENCES

[1]   Andrew Brown JD, Bernal JD. The sage of science. New York: Oxford University Press 2005; 343.

[2]   Lum KK, Cristea IM. Proteomic approaches to uncovering virus-host protein interactions during the progression of viral infection. Expert Rev Proteomics 2016; 13(3): 325-40.
[http://dx.doi.org/10.1586/14789450.2016.1147353] [PMID: 26817613]

[3]   Ideker T, Sharan R. Protein networks in disease. Genome Res 2008; 18(4): 644-52.
[http://dx.doi.org/10.1101/gr.071852.107] [PMID: 18381899]

[4]   Dyer MD, Neff C, Dufford M, *et al.* The human-bacterial pathogen protein interaction networks of Bacillus anthracis, Francisella tularensis, and Yersinia pestis. PLoS One 2010; 5(8): e12089.
[http://dx.doi.org/10.1371/journal.pone.0012089] [PMID: 20711500]

[5]   Malik-Soni N, Frappier L. Proteomic profiling of EBNA1-host protein interactions in latent and lytic Epstein-Barr virus infections. J Virol 2012; 86(12): 6999-7002. [published correction appears in J Virol. 2015 Sep;89(17):9143].
[http://dx.doi.org/10.1128/JVI.00194-12] [PMID: 22496234]

[6]   Diner BA, Lum KK, Javitt A, Cristea IM. Interactions of the Antiviral Factor Interferon Gamma-Inducible Protein 16 (IFI16) Mediate Immune Signaling and Herpes Simplex Virus-1 Immunosuppression. Mol Cell Proteomics 2015; 14(9): 2341-56.

[http://dx.doi.org/10.1074/mcp.M114.047068] [PMID: 25693804]

[7]   Bian X, Wiktor P, Kahn P, *et al.* Antiviral antibody profiling by high-density protein arrays. Proteomics 2015; 15(12): 2136-45.
[http://dx.doi.org/10.1002/pmic.201400612] [PMID: 25758251]

[8]   Yu X, Decker KB, Barker K, *et al.* Host-pathogen interaction profiling using self-assembling human protein arrays. J Proteome Res 2015; 14(4): 1920-36.
[http://dx.doi.org/10.1021/pr5013015] [PMID: 25739981]

[9]   Hall DA, Ptacek J, Snyder M. Protein microarray technology. Mech Ageing Dev 2007; 128(1): 161-7.
[http://dx.doi.org/10.1016/j.mad.2006.11.021] [PMID: 17126887]

[10]   Bantscheff M, Lemeer S, Savitski MM, Kuster B. Quantitative mass spectrometry in proteomics: critical review update from 2007 to the present. Anal Bioanal Chem 2012; 404(4): 939-65.
[http://dx.doi.org/10.1007/s00216-012-6203-4] [PMID: 22772140]

[11]   Sharma D, Singh R, Deo N, Bisht D. Interactome analysis of Rv0148 to predict potential targets and their pathways linked to aminoglycosides drug resistance: An insilico approach. Microb Pathog 2018; 121: 179-83.
[http://dx.doi.org/10.1016/j.micpath.2018.05.034] [PMID: 29800702]

[12]   Zhang QC, Petrey D, Deng L, *et al.* Structure-based prediction of protein-protein interactions on a genome-wide scale. Nature 2012; 490(7421): 556-60. [published correction appears in Nature. 2013 Mar 7;495(7439):127].
[http://dx.doi.org/10.1038/nature11503] [PMID: 23023127]

[13]   Zahiri J, Bozorgmehr JH, Masoudi-Nejad A. Computational prediction of protein-protein interaction networks: algo-rithms and resources. Curr Genomics 2013; 14(6): 397-414.
[http://dx.doi.org/10.2174/1389202911314060004] [PMID: 24396273]

[14]   Mewes HW, Frishman D, Güldener U, *et al.* MIPS: a database for genomes and protein sequences. Nucleic Acids Res 2002; 30(1): 31-4.
[http://dx.doi.org/10.1093/nar/30.1.31] [PMID: 11752246]

[15]   Bader GD, Betel D, Hogue CW. BIND: the Biomolecular Interaction Network Database. Nucleic Acids Res 2003; 31(1): 248-50.
[http://dx.doi.org/10.1093/nar/gkg056] [PMID: 12519993]

[16]   Salwinski L, Miller CS, Smith AJ, Pettit FK, Bowie JU, Eisenberg D. The Database of Interacting Proteins: 2004 update. Nucleic Acids Res 2004; 32(Database issue): D449-51.
[http://dx.doi.org/10.1093/nar/gkh086] [PMID: 14681454]

[17]   Chatr-Aryamontri A, Breitkreutz BJ, Oughtred R, *et al.* The BioGRID interaction database: 2015 update. Nucleic Acids Res 2015; 43(Database issue): D470-8.
[http://dx.doi.org/10.1093/nar/gku1204] [PMID: 25428363]

[18]   Hellman LM, Fried MG. Electrophoretic mobility shift assay (EMSA) for detecting protein-nucleic acid interactions. Nat Protoc 2007; 2(8): 1849-61.
[http://dx.doi.org/10.1038/nprot.2007.249] [PMID: 17703195]

[19]   Bazan J, Całkosiński I, Gamian A. Phage display--a powerful technique for immunotherapy: 1. Introduction and potential of therapeutic applications. Hum Vaccin Immunother 2012; 8(12): 1817-28.
[http://dx.doi.org/10.4161/hv.21703] [PMID: 22906939]

[20]   Jerabek-Willemsen M, Wienken CJ, Braun D, Baaske P, Duhr S. Molecular interaction studies using microscale thermophoresis. Assay Drug Dev Technol 2011; 9(4): 342-53.
[http://dx.doi.org/10.1089/adt.2011.0380] [PMID: 21812660]

[21]   Biomarkers Definitions Working Group.. Biomarkers and surrogate endpoints: preferred definitions and conceptual framework. Clin Pharmacol Ther 2001; 69(3): 89-95.
[http://dx.doi.org/10.1067/mcp.2001.113989] [PMID: 11240971]

[22]   Nelson MI, Holmes EC. The evolution of epidemic influenza. Nat Rev Genet 2007; 8(3): 196-205.
[http://dx.doi.org/10.1038/nrg2053] [PMID: 17262054]

[23]   Xu R, McBride R, Paulson JC, Basler CF, Wilson IA. Structure, receptor binding, and antigenicity of influenza virus hemagglutinins from the 1957 H2N2 pandemic. J Virol 2010; 84(4): 1715-21.
[http://dx.doi.org/10.1128/JVI.02162-09] [PMID: 20007271]

[24]   Taubenberger JK, Morens DM. 1918 Influenza: the mother of all pandemics. Emerg Infect Dis 2006; 12(1): 15-22.
[http://dx.doi.org/10.3201/eid1209.05-0979] [PMID: 16494711]

[25]   Shapira SD, Gat-Viks I, Shum BO, *et al.* A physical and regulatory map of host-influenza interactions reveals pathways in H1N1 infection. Cell 2009; 139(7): 1255-67.
[http://dx.doi.org/10.1016/j.cell.2009.12.018] [PMID: 20064372]

[26]   Tisoncik-Go J, Gasper DJ, Kyle JE, *et al.* Integrated Omics Analysis of Pathogenic Host Responses during Pandemic H1N1 Influenza Virus Infection: The Crucial Role of Lipid Metabolism. Cell Host Microbe 2016; 19(2): 254-66.
[http://dx.doi.org/10.1016/j.chom.2016.01.002] [PMID: 26867183]

[27]   Marion T, Elbahesh H, Thomas PG, DeVincenzo JP, Webby R, Schughart K. Respiratory Mucosal Proteome Quantification in Human Influenza Infections. PLoS One 2016; 11(4): e0153674.
[http://dx.doi.org/10.1371/journal.pone.0153674] [PMID: 27088501]

[28]   Alter MJ, Margolis HS, Krawczynski K, *et al.* The natural history of community-acquired hepatitis C in the United States. The Sentinel Counties Chronic non-A, non-B Hepatitis Study Team. N Engl J Med 1992; 327(27): 1899-905.
[http://dx.doi.org/10.1056/NEJM199212313272702] [PMID: 1280771]

[29]   Tsutsumi T, Matsuda M, Aizaki H, *et al.* Proteomics analysis of mitochondrial proteins reveals overexpression of a mitochondrial protein chaperon, prohibitin, in cells expressing hepatitis C virus core protein. Hepatology 2009; 50(2): 378-86.
[http://dx.doi.org/10.1002/hep.22998] [PMID: 19591124]

[30]   Kou YH, Chou SM, Wang YM, *et al.* Hepatitis C virus NS4A inhibits cap-dependent and the viral IRES-mediated translation through interacting with eukaryotic elongation factor 1A. J Biomed Sci 2006; 13(6): 861-74.
[http://dx.doi.org/10.1007/s11373-006-9104-8] [PMID: 16927014]

[31]   Albuquerque LM, Trugilho MR, Chapeaurouge A, *et al.* Two-dimensional difference gel electrophoresis (DiGE) analysis of plasmas from dengue fever patients. J Proteome Res 2009; 8(12): 5431-41.
[http://dx.doi.org/10.1021/pr900236f] [PMID: 19845402]

[32]   Higa LM, Caruso MB, Canellas F, *et al.* Secretome of HepG2 cells infected with dengue virus: implications for pathogenesis. Biochim Biophys Acta 2008; 1784(11): 1607-16.
[http://dx.doi.org/10.1016/j.bbapap.2008.06.015] [PMID: 18639654]

[33]   Ances BM, Ellis RJ. Dementia and neurocognitive disorders due to HIV-1 infection. Semin Neurol 2007; 27(1): 86-92.
[http://dx.doi.org/10.1055/s-2006-956759] [PMID: 17226745]

[34]   Rozek W, Ricardo-Dukelow M, Holloway S, *et al.* Cerebrospinal fluid proteomic profiling of HIV--infected patients with cognitive impairment. J Proteome Res 2007; 6(11): 4189-99.
[http://dx.doi.org/10.1021/pr070220c] [PMID: 17929958]

[35]   Shen B, Yi X, Sun Y, *et al.* Proteomic and Metabolomic Characterization of COVID-19 Patient Sera. Cell 2020; 182(1): 59-72.e15.
[http://dx.doi.org/10.1016/j.cell.2020.05.032] [PMID: 32492406]

[36]   Ihling C, Tänzler D, Hagemann S, *et al.* Mass Spectrometric Identification of SARS-CoV-2 Proteins from Gargle Solution Samples of COVID-19 Patients [published online ahead of print, 2020 Jun 23]. J

Proteome Res 2020.
[http://dx.doi.org/10.1021/acs.jproteome.0c00280]

[37]   Bojkova D, Klann K, Koch B, *et al.* Proteomics of SARS-CoV-2-infected host cells reveals therapy targets. Nature 2020; 583(7816): 469-72.
[http://dx.doi.org/10.1038/s41586-020-2332-7] [PMID: 32408336]

[38]   Ojima-Kato T, Yamamoto N, Nagai S, *et al.* Application of proteotyping Strain Solution™ ver. 2 software and theoretically calculated mass database in MALDI-TOF MS typing of *Salmonella* serotype. Appl Microbiol Biotechnol 2017; 101(23-24): 8557-69.
[http://dx.doi.org/10.1007/s00253-017-8563-3] [PMID: 29032472]

[39]   Leung LM, Fondrie WE, Doi Y, *et al.* Identification of the ESKAPE pathogens by mass spectrometric analysis of microbial membrane glycolipids. Sci Rep 2017; 7(1): 6403.
[http://dx.doi.org/10.1038/s41598-017-04793-4] [PMID: 28743946]

[40]   Patel TS, Kaakeh R, Nagel JL, Newton DW, Stevenson JG. Cost Analysis of Implementing Matrix-Assisted Laser Desorption Ionization-Time of Flight Mass Spectrometry Plus Real-Time Antimicrobial Stewardship Intervention for Bloodstream Infections. J Clin Microbiol 2016; 55(1): 60-7.
[http://dx.doi.org/10.1128/JCM.01452-16] [PMID: 27795335]

[41]   Perez KK, Olsen RJ, Musick WL, *et al.* Integrating rapid pathogen identification and antimicrobial stewardship significantly decreases hospital costs. Arch Pathol Lab Med 2013; 137(9): 1247-54.
[http://dx.doi.org/10.5858/arpa.2012-0651-OA] [PMID: 23216247]

[42]   Bush K. Proliferation and significance of clinically relevant β-lactamases. Ann N Y Acad Sci 2013; 1277: 84-90.
[http://dx.doi.org/10.1111/nyas.12023] [PMID: 23346859]

[43]   Schwarz S, Kehrenberg C, Doublet B, Cloeckaert A. Molecular basis of bacterial resistance to chloramphenicol and florfenicol. FEMS Microbiol Rev 2004; 28(5): 519-42.
[http://dx.doi.org/10.1016/j.femsre.2004.04.001] [PMID: 15539072]

[44]   Munita JM, Arias CA. Mechanisms of Antibiotic Resistance. Microbiol Spectr 2016; 4(2)
[http://dx.doi.org/10.1128/microbiolspec.VMBF-0016-2015]

[45]   Maxson T, Taylor-Howell CL, Minogue TD. Semi-quantitative MALDI-TOF for antimicrobial susceptibility testing in *Staphylococcus aureus*. PLoS One 2017; 12(8): e0183899.
[http://dx.doi.org/10.1371/journal.pone.0183899] [PMID: 28859120]

[46]   Ho PL, Yau CY, Ho LY, *et al.* Rapid detection of *cfiA* metallo-β-lactamase-producing *Bacteroides fragilis* by the combination of MALDI-TOF MS and CarbaNP. J Clin Pathol 2017; 70(10): 868-73.
[http://dx.doi.org/10.1136/jclinpath-2017-204335] [PMID: 28360188]

[47]   Klockgether J, Tümmler B. Recent advances in understanding *Pseudomonas aeruginosa* as a pathogen. F1000 Res 2017; 6: 1261.
[http://dx.doi.org/10.12688/f1000research.10506.1] [PMID: 28794863]

[48]   Hirsch EB, Tam VH. Impact of multidrug-resistant *Pseudomonas aeruginosa* infection on patient outcomes. Expert Rev Pharmacoecon Outcomes Res 2010; 10(4): 441-51.
[http://dx.doi.org/10.1586/erp.10.49] [PMID: 20715920]

[49]   Charretier Y, Köhler T, Cecchini T, *et al.* Label-free SRM-based relative quantification of antibiotic resistance mechanisms in *Pseudomonas aeruginosa* clinical isolates. Front Microbiol 2015; 6: 81.
[http://dx.doi.org/10.3389/fmicb.2015.00081] [PMID: 25713571]

[50]   de Keijzer J, Mulder A, de Ru AH, van Soolingen D, van Veelen PA. Parallel reaction monitoring of clinical *Mycobacterium tuberculosis* lineages reveals pre-existent markers of rifampicin tolerance in the emerging Beijing lineage. J Proteomics 2017; 150: 9-17.
[http://dx.doi.org/10.1016/j.jprot.2016.08.022] [PMID: 27576137]

[51]   Kruh-Garcia NA, Wolfe LM, Chaisson LH, *et al.* Detection of *Mycobacterium tuberculosis* peptides in

the exosomes of patients with active and latent *M. Tuberculosis* infection using MRM-MS. PLoS One 2014; 9(7): e103811.
[http://dx.doi.org/10.1371/journal.pone.0103811] [PMID: 25080351]

[52]　Kruh-Garcia NA, Murray M, Prucha JG, Dobos KM. Antigen 85 variation across lineages of *Mycobacterium tuberculosis*-implications for vaccine and biomarker success. J Proteomics 2014; 97: 141-50.
[http://dx.doi.org/10.1016/j.jprot.2013.07.005] [PMID: 23891556]

[53]　Larson MA, Ding SJ, Slater SR, *et al.* Application of chromosomal DNA and protein targeting for the identification of Yersinia pestis. Proteomics Clin Appl 2013; 7(5-6): 416-23.
[http://dx.doi.org/10.1002/prca.201200092] [PMID: 23436733]

CHAPTER 9

Microbial Metalloproteome: Approaches and Biomedical Application in Microbial Antibiotics Resistance

Saroj Sharma, **Monalisa Tiwari** and **Vishvanath Tiwari**[*]

Department of Biochemistry, Central University of Rajasthan, Ajmer-305817, India

Abstract: Microbial metalloproteomics involves a detailed analysis of the proteins that have metals as an important part or known to bind to metals in biological samples. Recent updates showed that metalloproteome helps in understanding the role of different environments/conditions in the survival of microbes and is involved in microbial pathogenesis. Microbial metalloproteomics could also be used in understanding the resistance mechanisms of microbes. We have explored different metalloproteomic approaches such as inductively coupled plasma-Mass spectroscopy (ICP)-MS, X-ray absorption/fluorescence, radionuclide, and bioinformatics. We have also discussed the role of metalloproteins such as metallo-beta-lactamases in microbial drug resistance, the alternation of the microbial proteome in response to the metal, and their role in host-pathogen interactions. We have also surveyed different therapeutics targeting the microbial metalloproteins. Current advancements in the metalloproteome would help in understanding the mechanism better and the adequate role of metalloenzymes and proteins in conferring the drug resistance in microbes.

Keywords: Host-pathogen interaction, ICP-MS, Metalloproteome, Metallo-beta lactamases, Metalloproteomic approaches, Metal inducible proteome, Microbial drug resistance, Microbial therapeutics, X-ray absorption.

INTRODUCTION

Metalloproteome is a very common term used for a protein that contains metal ions. Metals form a very important part of our biological system, and their incorporation into the protein ensures the proper functioning of the enzyme. For example, Zinc containing proteins have an important role in the biological world, including fixation and synthesis of Carbon, and regulation of redox reactions [1]. Zinc is the most abundant part of the biological world as it represents 10% of the

[*] **Corresponding author Vishvanath Tiwari:** Department of Biochemistry, Central University of Rajasthan, Ajmer-305817, India; Tel: +91-850-300-2573; E-mail: vishvanath@curaj.ac.in

Divakar Sharma (Ed.)
All rights reserved-© 2020 Bentham Science Publishers

entire proteome, whereas the cellular zinc content remains in submillimolar ranges [2]. Metalloproteins have various roles in the cell, like storage and transport of proteins, enzymes, and signal transduction [3]. Studies have projected that about 30% of protein candidates are known to have metal ions as cofactors [4]. This field has started gaining importance in the current scenario, such as the role of metalloproteins has also been documented in Alzheimer's disease [5] and it helps in predicting the function of metal ions in the molecular mechanism of diseases. Different metals are required for diverse biological functions (Table **1**). Metalloproteomics is an area that includes various approaches covering the expression of metalloproteins and their variations in a biological system [6]. The field of metalloproteomics offers novel insights into the basic biological processes. To analyze the potential of this field, it is very important to look forward to the methodological advances [7].

Table 1. Biological function of various metal ions that are found abundantly in the living system. (Chemistry of the Elements, Greenwood 2004).

Name of the Metal	The Biological Function of the Metal
Zn	A part of metalloenzymes like carbonic anhydrase, Calcium ion absorption, and structure maintenance
Cu	Oxygen transport, electron transfer, redox reaction catalysis
Ni	Hydrogenase, hydrolase, iron absorption
Co	Development and maintenance of blood vessels, skin, bones and joints, Redox catalysis
Fe	Redox catalysis, oxygen transport and electron transport
Mn	Have antioxidant property as it is a component of Superoxide dismutase, Nitrogen fixation
Mo	Nitrogen fixation, oxo transfer, electron transfer
Mg	Protein synthesis, muscle and nerve, oxidative phosphorylation, and glycolysis
Ca	Cell signaling, structure, and carrier of charge

EMERGENCE OF METALLOPROTEOMICS

Metals administered in our body unnaturally (through environmental contamination/ with drugs) are known to have various effects on the body. The varying coordination numbers, sites, and solvent accessibilities of these sites ensure the current placement of metals in terms of location, concentration, and time of delivery for human health [8]. One of the most common examples is zinc (Zn(II)), representing 10% of the proteome [2]. Zinc has an important function in various pathways, like fixation of carbon, biosynthesis, and redox regulation. Though Zinc is an essential metal, overloaded amounts of Zn(II) are toxic for the body because it competes for binding with the metalloproteins destined for Fe(II),

Mn(II) and Co(II) [3]. To overcome this problem, the cell over-expresses zinc-containing proteins to manage the cytosolic concentration of zinc in the free form [8]. This has led to a hypothesis that as soon as the zinc ion enters the cells, it changes the sites of presence, and in the precisely same manner, copper shifts its position concerning the binding efficiencies in the cytosol [9]. A complex network of interacting proteins called "interactomes" manages the zinc distribution in the sites that are known to compete with each other [10]. But the pathway of the zinc ions to reach their targets is still undiscovered. The uptake and efflux of zinc (zinc interactome) are clear [11]; beyond this point, the pathway of zinc transport is entirely unknown. For example, the secretory granules of pancreatic β-cells store insulin in the form of zinc-coordinated hexamers [12]. The abnormal insulin secretion due to the impairments of β-cells is the crucial factor for the development of glucose tolerance to the disease type-2 diabetes [10]. Another example is the entrapment of zinc in senile plaques during Alzheimer's [13]. The plaque formation takes place due to the aggregation of amyloid-beta peptides (Aβ), and Zinc facilitates the seeding by binding to Aβ [14]. Tracing the path utilised by Zinc to finds its way to Aβ can reveal information about the neural pathogenesis of Alzheimer's disease; hence suggesting new targets to develop drugs [10]. For a better understanding of the working of these metals, we need new techniques. Imaging is one of the reliable tools that provide information about the concentration and localization of various minerals in the cells. The development of selective metal-ion sensors with optical-fluorescence is one of the primary examples. Hence new fluorescent indicators have been synthesized by measuring the concentration of free Ca^{2+} in the cytosol [15]. Detectors for Cu, Zn, Fe are based on the usage of photochemical tools like fluorescent sensors with photocaged complexes to look into the details of homeostasis and signalling mechanism [16]. X-ray fluorescence microscopy (XFM), due to the deep penetrating activity of X-rays, shows marvellous trace element sensitivity [17]. The specimens can be observed in their natural state as there is no requirement of sectioning in XFM.

Among the techniques developed for the imaging of metals at the cellular level is the coupling of liquid chromatography with inductively-coupled-plasma mass-spectrometry (ICP-MS). ICP-MS has been used as a detector in chromatography and electrophoresis. It is supported by electrospray and Matrix-assisted laser desorption/ionisation MALDI MS, being one of the top analytical techniques [18]. Many other approaches are yet to be explored, and the current existing has its advantages and limitations.

CURRENT APPROACHES TO METALLOPROTEOMICS

Almost one-third of the proteins have metal ions in conjugated form, the most abundant being magnesium and Zinc. There have been different approaches associated with metalloproteomics. It includes both forward (bottom-up) and reverse (top-down) technologies (Fig. **1**). In reverse technology, the samples are fractionated in a sequence-wise chromatographic or 2-D gel separation. Those fractionated samples are subjected to atomic absorption and mass spectrometry (LC-MS and/or ICP-MS) for the identification of metal-associated specific proteins [19]. The isolation of the proteins from its native cell/tissue environment is one of the main advantages, while isolation of proteins of interest and loss of native metals are one of the main disadvantages. Tainer JA and Adams MWW *et al.* have reported such work in *Pyrococcus furiosus*, *Escherichia coli*, and *Sulfolobus solfataricus* on genome-scale [20]. The analysis of three hundred forty-three metal peaks by ICP-MS in *Pyrococcus furiosus* identified iron (Fe), Zinc (Zn), tungsten (W) and nickel (Ni) and cobalt (Co) as the most abundant metals [21]. Matrix-assisted laser desorption/ionization (MALDI)-MS also showed the presence of various Nickel- and molybdenum-containing proteins.

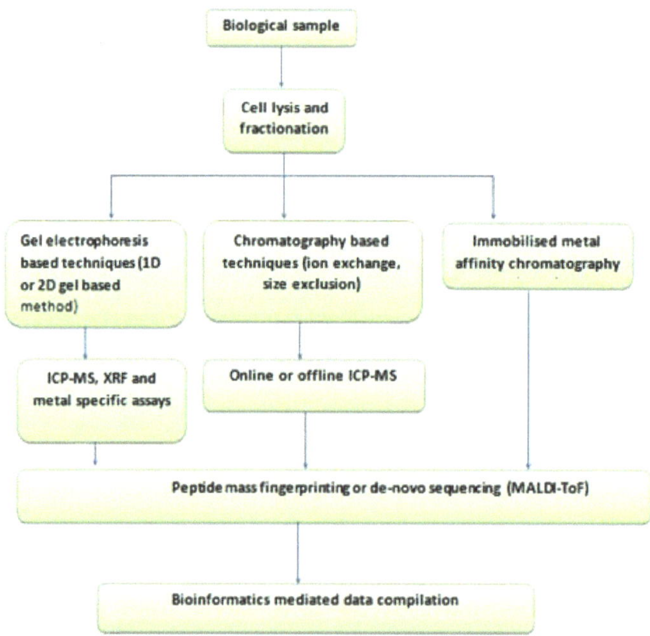

Fig. (1). Flow chart representing the basic methodologies in metalloproteomics.

The metalloproteins are fractionated based on the differential binding affinity between amino acids that are exposed on the surface by using Immobilized-metal affinity chromatography (IMAC) [22]. The samples that are devoid of metals are loaded upon IMAC chips that have already been saturated with known metals. Hence the proteins having an affinity with that particular metal are recovered and the detailed analysis is performed using 2-D gels or surface-enhanced laser desorption ionization (SELDI) MS [23]. This also targets the post-translation modification of proteins as it involves metalloenzymes, metal transporters, and metallochaperones. The phosphoproteome in *Streptococcus pneumonia* was identified by combining IMAC and the hybrid LTQ-orbitrap MS [24]. This technique is unable to identify the metal-protein complexes and metalloproteins that have a high binding affinity between the metal and the proteins as they pass undetected through the chips.

On the other hand, cloning and expression of genes of interest which is followed by investigating the metal content, are one of the main aspects of the forward (bottom-up) approach. The strength of this method lies in the ability to control the expression of the protein of interest. Whereas the loss of native metals during purification steps is the main loophole that needs to be mend [25].

The reverse and the forward strategies both show the mis-annotation of a metalloprotein as there is always a chance of losing the metals during purification. Another drawback is the false-negative results accompanying unassociated metals with the proteins. The presence of non-native metallic ions could also confuse the investigators about the functionality of the proteins. One has to be extremely careful while handling such experimental processes. One can rely on the use of computational methods as several bioinformatic tools could predict the metal-binding proteins based on consensus sequences and the 3-D structure of a protein [26]. The complete metalloproteins are considered in this approach; hence these could be applied to the whole-genome sequences as well.

The field of metalloproteomics is also known to involve the functional and structural characterization of metal-binding proteins. X-ray absorption spectroscopy (XAS), X-ray solution scattering (SAXS), and X-ray macromolecular crystallography (MX) are a part of synchrotron radiation which is involved in the structural and functional studies of metalloprotein complexes. The proteins that are usually missed in the medium-low resolution structures are generally recovered by the atomic resolution [27]. The active sites could be predicted with extreme accuracies using XAS and the data for thousands of protein samples could be collected in a single run. The non-crystallisable protein samples could be analyzed here to know about the element related to the structure of various reaction intermediates. On the other hand, SAXS has a relatively low resolution (~15 Å) for the determination of protein-protein complexes for resolving the biological function of proteins [28]. The HT-XAS (high- throughput

XAS) uses the basic principle of excitation of electrons of the metal atoms in combination with the detection of fluorescence in a way similar to the X-ray absorption spectroscopy [21]. Hence to understand the metal-binding sites, one can effectively combine both the experimental as well as bioinformatic approaches.

ROLE OF METALLOPROTEINS IN MICROBIAL DRUG RESISTANCE

The microbes are smart enough to search for survival strategies against the drugs, antibiotics. β-lactams are the most common antibiotics used, have a four-membered cyclic amide ring, and they copy the mechanism of cleavage of terminal D-Ala-D-Ala bond cleaved by transpeptidases during cross-linking of the peptidoglycans [29]. The cell wall is weakened due to the inability to displace the ester intermediate as the β-lactam bond is attacked by the same active-site serine nucleophile by binding to the active sites of transpeptidases [30]. With the increase in using penicillin to target the microbes and development of derivatives of penicillin (penam, clavam, and carbapenem), various bacteria have developed multiple resistance mechanisms [31], which include expression of the β-lactamase enzyme (zinc-containing enzyme). The most widely used class of β-lactamase is Class B β-lactamase, which provides imipenemase (IMP).

β-Lactamases hydrolyze the β-lactam amide bond and convert the ring structures into open products that are no longer capable of transpeptidase inhibition. A new class of β-lactamase that was chromosomally encoded in *Bacillus cereus* and was inhibited on the addition of the metal ion chelatorethylenediaminetetraacetic acid (EDTA), indicated the first discovery of a metal-dependent β-lactamase [32]. This led to the development of more resistant strains such as CcrA (carbapenem and cephamycin resistance A) from *Bacteroides fragilis*, CphA from *Aeromonas hydrophila,* and the first transferable MBL, IMP-1 from *Pseudomonas aeruginosa* [33]. Sequences encoding IMP-type MBLs are commonly found in *P. aeruginosa* (mainly belonging to the clonal complex, CC111, and CC235), *Acinetobacter baumannii*, and *Enterobacteriaceae* isolates. However, they have also been occasionally identified in other organisms (*e.g., P. putida*). A second major subgroup in the B1 subclass is the VIM-type MBLs; the genes for VIM-type MBLs are strongly associated with class I integrons and integrated within either chromosomes or plasmids. The VIM-type MBLs are more reported in *Enterobacteriaceae and Pseudomonas aeruginosa*. A third major type of B1 β-lactamases is the NDMs; the gene for this has been reported in *Enterobacteriaceae* species and other Gram-negative bacteria, such as *Vibrio cholerae, Pseudomonas spp.* and *A. baumannii* [31].

METAL INDUCIBLE MICROBIAL PROTEOME AND HOST-PATHOGEN INTERACTION

This section covers the role of the metal ions (Fe, Cu, Zn) in infectious diseases. As soon as a microbial pathogen enters the host system, it starts competing with the host for nutrients. Iron is an essential nutrient for both mammals and microorganisms as it exerts subtle effects on the proliferation and functionality of immune cells. It is a co-factor for proteins involved in metabolism, DNA synthesis, and electron transfer. The control over iron homeostasis is thus of vital significance in host-pathogen interaction [34]. Extracellular bacteria such as *Staphylococcus aureus* damage erythrocytes by toxic hemolysins to use these cells as iron sources, while intracellular pathogens, including protozoa, acquire Hb or haem molecules from cell surface receptors [35]. *Leishmania* lack the pathway for haem biosynthesis, so they express specific Hb and haem receptors because haem acquisition is of pivotal importance for the survival of these parasites [36]. *Trypanosoma* species can utilize both haem and ionic iron. *Plasmodium falciparum* utilizes Hb and haem within RBCs and avoids haem-mediated oxidative damage by metabolizing it either to haemozoin or biliverdin [37]. *Mycobacterium leprae* uptakes Cu^{2+} and Zn^{2+} essential for the generation of superoxide dismutases and catalases, responsible for providing defense against reactive oxygen species based death in macrophages [38]. Recent studies have revealed that the most probable site for Zn^{2+}-binding lies in the N-terminal domain; while, the same for Cu^{2+}-binding lies in the "alpha-crystallin domain" *Mycobacterium leprae* HSP18.

TARGETING METALLOPROTEINS TO TARGET MICROBE

As the host is infected by a microbial pathogen, the host organism sequesters trace minerals to limit the pathogenicity, a process called "nutritional immunity". For example, siderophores (iron-chelators) are secreted by pathogens and chelate iron with a higher affinity than transferrin [39]. Pathogens then utilize Siderophore-iron complexes through cell surface receptors, and the bound iron is released through ferric reductases, reducing ferric (Fe^{3+}) to the soluble ferrous (Fe^{2+}). To overcome this problem, hosts produce siderocalin that binds and sequesters siderophores, preventing their uptake by pathogens. Zinc uptake by pathogens occurs *via* zinc-binding siderophores (zincophores) [40]. *Yersinia pestis,* which causes bubonic plague, utilizes Ybt as a zincophore and ZnuABC to acquire Zinc and cause lethal infection in a septicemic plague mouse. Hence the host needs to create hypo-metallic environments to reduce the pathogenicity of microbes.

Another approach to targeting metalloproteins for reducing the chances of infection by microbes is β-lactamase based drugs. Beta-lactamase inhibitors are a class of compounds that block the activity of beta-lactamase enzymes (also called beta-lactamases), preventing the degradation of beta-lactam antibiotics. A β-lactamase inhibitor isolated from *Streptomyces clavuligerus* is termed clavulanic acid. It is a potent inhibitor of many β-lactamases, including the ones of *Escherichia coli* (plasmid-mediated), *Klebsiellaaerogenes, Proteus mirabilis*, and *Staphylococcus aureus* [41]. Clavulanic acid covalently binds to a serine residue at the active site of the β-lactamase. This restructures the clavulanic acid molecule, creating a much more reactive species that attacks another amino acid in the active site, permanently inactivating it, and thus inactivating the enzyme. Another β-lactamase inhibitor, sulbactam, can inhibit the most common forms of β-lactamase. It confers little protection against bacteria such as *Pseudomonas aeruginosa, Citrobacter, Enterobacter*, and *Serratia* as it is not able to interact with the AmpC cephalosporinase [42]. Tazobactam (another potent inhibitor of β-lactamase) is combined with the extended spectrum β-lactam antibiotic piperacillin in the drug piperacillin/tazobactam, which is used in infections caused by *Pseudomonas aeruginosa*. Hence the menace of multidrug resistance can be overcome by using the above-mentioned drugs and by combining them with the pre-existing antibiotics.

CONCLUDING REMARKS AND FUTURE ASPECTS

Metals have a very important role in biochemistry, such as the importance of minerals in human diseases, *e.g.*, in Alzheimer's and in infectious diseases that have started gaining considerable attention in recent years. The two approaches, bottom-up metalloproteomics, are used to characterize an organism as a whole (complete genome) and provide information on the possible metalloproteome [43]. One of the main limitations is that not all metalloproteins are encoded at a given time and growth conditions, whereas top-down approaches study the microbial metalloproteome under different external conditions but are limited by the metal concentration and protein analysis techniques. Culminating computational predictions with experimental metalloproteomics will lead to an advanced metalloproteomic workflow that can provide essential information to complete the models in systems biology and to improve the designs in synthetic biology. Absolute quantification of purified protein is direct, whereas the absolute quantification of a single protein in a complex mixture using MS can be readily achieved by using reference peptides. Still, there are a large number of proteins that are mysterious in proteomics. Metal/protein stoichiometry can be determined by using the absolute metal quantity and the absolute protein quantity, which can lead to the identification of new proteins.

There are various modern imaging techniques like nanoSIMS (Secondary Ion Mass Spectrometry), scanning electron microscopy with EDS (Energy Dispersive X-ray Spectroscopy), and EFTEM (Energy Filtering Transmission Electron Microscopy) in principle that allow imaging of metals in individual cells, even bacteria [44]. The fluorescent sensors are used to measure "free" metal ions in living cells as these have been successfully reported for Ca^{2+}, Cu^{2+}, and Zn^{2+}. For example, Genetically-encoded FRET sensors detect the presence of free Zn^{2+} in living cells [45]. Although this does not provide metalloproteome information *per se*, the data can be valuable to study the metallome, which also includes metals not bound to proteins. The ability to measure metals ions in a living cell is a crucial property of the modern surge in super-resolution fluorescence microscopy. Moreover, it is worth noting here that, using fluorescence microscopy, it is possible to study metals in micro-organisms. With the day to day advancements in the field of technology, many scopes are still unexplored in the field of metalloproteomics.

CONSENT FOR PUBLICATION

Not applicable.

CONFLICT OF INTEREST

The authors declare no conflict of interest, financial or otherwise.

ACKNOWLEDGEMENTS

SS and MT would like to thank the Central University of Rajasthan for their Ph.D. fellowship.

REFERENCES

[1]   Silva CL, Williams RJP. The biological chemistry of the elements: the inorganic chemistry of life. Oxford: Oxford University Press 2009.

[2]   Dupont CL, Yang S, Palenik B, Bourne PE. Modern proteomes contain putative imprints of ancient shifts in trace metal geochemistry. Proc Natl Acad Sci USA 2006; 103(47): 17822-7.
[http://dx.doi.org/10.1073/pnas.0605798103] [PMID: 17098870]

[3]   Waldron KJ, Robinson NJ. How do bacterial cells ensure that metalloproteins get the correct metal? Nat Rev Microbiol 2009; 7(1): 25-35.
[http://dx.doi.org/10.1038/nrmicro2057] [PMID: 19079350]

[4]   Tainer JA, Roberts VA, Getzoff ED. Protein metal-binding sites. Curr Opin Biotechnol 1992; 3(4): 378-87.
[http://dx.doi.org/10.1016/0958-1669(92)90166-G] [PMID: 1368439]

[5]   Hare DJ, Rembach A, Roberts BR. The emerging role of metalloproteomics in alzheimer's disease research. Methods Mol Biol 2016; 1303: 379-89.
[http://dx.doi.org/10.1007/978-1-4939-2627-5_22] [PMID: 26235079]

[6]   Maret W. Metalloproteomics, metalloproteomes, and the annotation of metalloproteins. Metallomics 2010; 2(2): 117-25.
[http://dx.doi.org/10.1039/B915804A] [PMID: 21069142]

[7]   Barnett JP, Scanlan DJ, Blindauer CA. Protein fractionation and detection for metalloproteomics: challenges and approaches. Anal Bioanal Chem 2012; 402(10): 3311-22.
[http://dx.doi.org/10.1007/s00216-012-5743-y] [PMID: 22302168]

[8]   Finney LA, O'Halloran TV. Transition metal speciation in the cell: insights from the chemistry of metal ion receptors. Science 2003; 300(5621): 931-6.
[http://dx.doi.org/10.1126/science.1085049] [PMID: 12738850]

[9]   Banci L, Bertini I, Ciofi-Baffoni S, Kozyreva T, Zovo K, Palumaa P. Affinity gradients drive copper to cellular destinations. Nature 2010; 465(7298): 645-8.
[http://dx.doi.org/10.1038/nature09018] [PMID: 20463663]

[10]   Lothian A, Hare DJ, Ryan TM, Masters CL, Roberts BR, Grimm R. Metalloproteomics: principles, challenges, and applications to neurodegeneration. Front Aging Neurosci Frontiers in Aging Neuroscience. 2013; 5: p. 35.

[11]   Nies DH. Biochemistry. How cells control zinc homeostasis. Science 2007; 317(5845): 1695-6.
[http://dx.doi.org/10.1126/science.1149048] [PMID: 17885121]

[12]   Adams MJ, Blundell TL, Dodson EJ, Dodson GG, Vijayan M, Baker EN, *et al.* Structure of rhombohedral 2 zinc insulin crystals. Nature 1969; 224(5218): 491-5.
[http://dx.doi.org/10.1038/224491a0]

[13]   Lovell MA, Robertson JD, Teesdale WJ, Campbell JL, Markesbery WR. Copper, iron and zinc in Alzheimer's disease senile plaques. J Neurol Sci 1998; 158(1): 47-52.
[http://dx.doi.org/10.1016/S0022-510X(98)00092-6] [PMID: 9667777]

[14]   Bush AI, Pettingell WH, Multhaup G, *et al.* Rapid induction of Alzheimer A beta amyloid formation by zinc. Science 1994; 265(5177): 1464-7.
[http://dx.doi.org/10.1126/science.8073293] [PMID: 8073293]

[15]   Minta A, Kao JP, Tsien RY. Fluorescent indicators for cytosolic calcium based on rhodamine and fluorescein chromophores. J Biol Chem 1989; 264(14): 8171-8.
[PMID: 2498308]

[16]   Mbatia HW, Burdette SC. Photochemical tools for studying metal ion signaling and homeostasis. Biochemistry 2012; 51(37): 7212-24.
[http://dx.doi.org/10.1021/bi3001769] [PMID: 22897393]

[17]   Paunesku T, Vogt S, Maser J, Lai B, Woloschak G. X-ray fluorescence microprobe imaging in biology and medicine. J Cell Biochem 2006; 99(6): 1489-502.
[http://dx.doi.org/10.1002/jcb.21047] [PMID: 17006954]

[18]   Szpunar J. Advances in analytical methodology for bioinorganic speciation analysis: metallomics, metalloproteomics and heteroatom-tagged proteomics and metabolomics. Analyst (Lond) 2005; 130(4): 442-65.
[http://dx.doi.org/10.1039/b418265k] [PMID: 15776152]

[19]   Manley SA, Byrns S, Lyon AW, Brown P, Gailer J. Simultaneous Cu-, Fe-, and Zn-specific detection of metalloproteins contained in rabbit plasma by size-exclusion chromatography-inductively coupled plasma atomic emission spectroscopy. J Biol Inorg Chem 2009; 14(1): 61-74.
[http://dx.doi.org/10.1007/s00775-008-0424-1] [PMID: 18781345]

[20]   Cvetkovic A, Menon AL, Thorgersen MP, *et al.* Microbial metalloproteomes are largely uncharacterized. Nature 2010; 466(7307): 779-82.
[http://dx.doi.org/10.1038/nature09265] [PMID: 20639861]

[21]   Shi W, Chance MR. Metalloproteomics: forward and reverse approaches in metalloprotein structural

and functional characterization. Curr Opin Chem Biol 2011; 15(1): 144-8.
[http://dx.doi.org/10.1016/j.cbpa.2010.11.004] [PMID: 21130021]

[22]　Sun X, Chiu J-F, He Q-Y. Fractionation of proteins by immobilized metal affinity chromatography 2008.

[23]　Roelofsen H, Balgobind R, Vonk RJ. Proteomic analyzes of copper metabolism in an *in vitro* model of Wilson disease using surface enhanced laser desorption/ionization-time of flight-mass spectrometry. J Cell Biochem 2004; 93(4): 732-40.
[http://dx.doi.org/10.1002/jcb.20226] [PMID: 15660417]

[24]　Sun X, Ge F, Xiao C-L, *et al.* Phosphoproteomic analysis reveals the multiple roles of phosphorylation in pathogenic bacterium Streptococcus pneumoniae. J Proteome Res 2010; 9(1): 275-82.
[http://dx.doi.org/10.1021/pr900612v] [PMID: 19894762]

[25]　Scott RA, Shokes JE, Cosper NJ, Jenney FE, Adams MW. Bottlenecks and roadblocks in high-throughput XAS for structural genomics. J Synchrotron Radiat 2005; 12(Pt 1): 19-22.
[http://dx.doi.org/10.1107/S0909049504028791] [PMID: 15616360]

[26]　Andreini C, Bertini I, Rosato A. Metalloproteomes: a bioinformatic approach. Acc Chem Res 2009; 42(10): 1471-9.
[http://dx.doi.org/10.1021/ar900015x] [PMID: 19697929]

[27]　Hasnain SS. Synchrotron techniques for metalloproteins and human disease in post genome era. J Synchrotron Radiat 2004; 11(Pt 1): 7-11.
[http://dx.doi.org/10.1107/S0909049503024166] [PMID: 14646121]

[28]　Hura GL, Menon AL, Hammel M, *et al.* Robust, high-throughput solution structural analyses by small angle X-ray scattering (SAXS). Nat Methods 2009; 6(8): 606-12.
[http://dx.doi.org/10.1038/nmeth.1353] [PMID: 19620974]

[29]　Tipper DJ, Strominger JL. Mechanism of action of penicillins: a proposal based on their structural similarity to acyl-D-alanyl-D-alanine. Proc Natl Acad Sci USA 1965; 54(4): 1133-41.
[http://dx.doi.org/10.1073/pnas.54.4.1133] [PMID: 5219821]

[30]　Tipper DJ. Mode of action of beta-lactam antibiotics. Rev Infect Dis 1979; 1(1): 39-54.
[http://dx.doi.org/10.1093/clinids/1.1.39] [PMID: 400939]

[31]　Mojica MF, Bonomo RA, Fast W. B1-metallo-β-lactamases: where do we stand? Curr Drug Targets 2016; 17(9): 1029-50.
[http://dx.doi.org/10.2174/1389450116666151001105622] [PMID: 26424398]

[32]　Sabath LDA, Abraham EP. Zinc as a cofactor for cephalosporinase from *Bacillus cereus* 569. Biochem J 1966; 98(1): 11C-3C.
[http://dx.doi.org/10.1042/bj0980011C] [PMID: 4957174]

[33]　Watanabe M, Iyobe S, Inoue M, Mitsuhashi S. Transferable imipenem resistance in *Pseudomonas aeruginosa.* Antimicrob Agents Chemother 1991; 35(1): 147-51.
[http://dx.doi.org/10.1128/AAC.35.1.147] [PMID: 1901695]

[34]　Nairz M, Schroll A, Sonnweber T, Weiss G. The struggle for iron - a metal at the host-pathogen interface. Cell Microbiol 2010; 12(12): 1691-702.
[http://dx.doi.org/10.1111/j.1462-5822.2010.01529.x] [PMID: 20964797]

[35]　Schaible UE, Collins HL, Priem F, Kaufmann SH. Correction of the iron overload defect in beta--microglobulin knockout mice by lactoferrin abolishes their increased susceptibility to tuberculosis. J Exp Med 2002; 196(11): 1507-13.
[http://dx.doi.org/10.1084/jem.20020897] [PMID: 12461085]

[36]　Sengupta S, Tripathi J, Tandon R, *et al.* Hemoglobin endocytosis in Leishmania is mediated through a 46-kDa protein located in the flagellar pocket. J Biol Chem 1999; 274(5): 2758-65.
[http://dx.doi.org/10.1074/jbc.274.5.2758] [PMID: 9915807]

[37]   Francis SE, Sullivan DJ, Goldberg Daniel E. Hemoglobin metabolism in the malaria parasite plasmodium falciparum. Annu Rev Microbiol 1997; 51(1): 97-123.
[http://dx.doi.org/10.1146/annurev.micro.51.1.97] [PMID: 9343345]

[38]   Nandi SK, Chakraborty A, Panda AK, Kar RK, Bhunia A, Biswas A. Evidences for zinc (II) and copper (II) ion interactions with Mycobacterium leprae HSP18: Effect on its structure and chaperone function. J Inorg Biochem 2018; 188: 62-75.
[http://dx.doi.org/10.1016/j.jinorgbio.2018.08.010] [PMID: 30121399]

[39]   Holden VI, Bachman MA. Diverging roles of bacterial siderophores during infection. Metallomics 2015; 7(6): 986-95.
[http://dx.doi.org/10.1039/C4MT00333K] [PMID: 25745886]

[40]   Hood MI, Skaar EP. Nutritional immunity: transition metals at the pathogen-host interface. Nat Rev Microbiol 2012; 10(8): 525-37.
[http://dx.doi.org/10.1038/nrmicro2836] [PMID: 22796883]

[41]   Reading C, Cole M. Clavulanic acid: a beta-lactamase-inhiting beta-lactam from Streptomyces clavuligerus. Antimicrob Agents Chemother 1977; 11(5): 852-7.
[http://dx.doi.org/10.1128/AAC.11.5.852] [PMID: 879738]

[42]   Totir MA, Helfand MS, Carey MP, *et al.* Sulbactam forms only minimal amounts of irreversible acrylate-enzyme with SHV-1 β-lactamase. Biochemistry 2007; 46(31): 8980-7.
[http://dx.doi.org/10.1021/bi7006146] [PMID: 17630699]

[43]   Hagedoorn P-L. Microbial Metalloproteomics. Proteomes 2015; 3(4): 424-39.
[http://dx.doi.org/10.3390/proteomes3040424] [PMID: 28248278]

[44]   Duhutrel P, Bordat C, Wu TD, Zagorec M, Guerquin-Kern JL, Champomier-Vergès MC. Iron sources used by the nonpathogenic lactic acid bacterium Lactobacillus sakei as revealed by electron energy loss spectroscopy and secondary-ion mass spectrometry. Appl Environ Microbiol 2010; 76(2): 560-5.
[http://dx.doi.org/10.1128/AEM.02205-09] [PMID: 19933352]

[45]   Hessels AM, Merkx M. Genetically-encoded FRET-based sensors for monitoring Zn(2+) in living cells. Metallomics 2015; 7(2): 258-66.
[http://dx.doi.org/10.1039/C4MT00179F] [PMID: 25156481]

CHAPTER 10

Proteomics of *Mycobacterium Tuberculosis*: An Overview

Anil Kumar Gupta[1,3], Divakar Sharma[2] and Amit Singh[1,3,*]

[1] *Department of Microbiology, All India Institute of Medical Sciences, Saket Nagar, Bhopal, India*

[2] *CRF, Kusuma School of Biological Sciences, Indian Institute of Technology, Delhi, India*

[3] *All India Institute of Medical Sciences, New Delhi, India*

Abstract: Tuberculosis (TB) is a universally prevalent disease caused by an aerobic, gram-positive bacterium *Mycobacterium tuberculosis* (*M. tuberculosis*). It has continued to pose a significant threat to human health. The emergence of multi-drug resistance (MDR) strains of *Mycobacterium tuberculosis* (*M. tuberculosis*) has further worsened the situation worldwide. Its genome has been primarily focusing on the past exploration of the molecular basis of disease. The genetic information or genes are transcribed into mRNA that is then processed, spliced, and translated into a single or multitude of proteins. Proteomics is the large-scale study of proteins, focusing on their structure and functions. For understanding the biology of any living organisms including humans, we need to decode the information encoded by proteins or its associated genes, as it reflects the true status of the cell. Although proteins play a major role in the whole life of the organism, its profile may vary from cell to cell due to spontaneous changes or biochemical interaction of microbial genome with environments. Even an immense amount of DNA/gene sequences data has been deposited by the scientific community in the databases, which is very useful in determining the virulencity, pathogenicity of the organisms. It simply contains a complete sequence of genomes that lacks its usefulness to illuminate biological function. The regulation process of a single cell involves complex mechanisms, including a multitude of metabolic and regulatory pathways for its survival. To date, no strict linear relationship has been documented between genes and proteome of the cell. The available technologies, *i.e.* microarray, identify a large number of differentially expressed genes very quickly. However, these have failed to identify the functional significance of the associated genes. In this chapter, we highlight prospects of the advances in the proteomics study that could be beneficial to mycobacterial research. Also, it will provide information to explore the possibility of protein-based biomarkers in the development of new diagnostic therapeutics for TB.

---

* **Corresponding author Amit Singh:** Department of Microbiology, All India Institute of Medical Sciences, Saket Nagar, Bhopal, India and All India Institute of Medical Sciences, New Delhi, India; Tel: 011-26549372; E-mail: amit4us@gmail.com

Divakar Sharma (Ed.)
All rights reserved-© 2020 Bentham Science Publishers

Keywords: Biomarkers, Drug discovery, Drug targets, *M. tuberculosis*, Metabolic pathways, Multidrug-resistant TB (MDR-TB), Proteomics, Structural biology, Structural genomics, Vaccine, XDR-TB.

INTRODUCTION

The "genome" has been a primary focus of the past exploration of the molecular basis of disease. The genetic information coded in genes is first transcribed into mRNA, which is then further processed, spliced, and translated into a multitude of proteins by strict complex mechanisms. Proteomics is a large-scale study of proteins, particularly their structure and functions. This term is coined to make an analogy with genomics. To understand the biology of any living things present on the planet, we need to know about macromolecules such as proteins that are the functional unit of the cell. Thus, proteins reflect the true status of the cell, and its differential expression may rule out the health status of the organisms. The proteome profile varies from cell to cell due to spontaneous changes or biochemical interaction of microbial genome with environmental conditions. Proteins play a central role in the life of an organism. The regulation process of a single cell involves complex mechanisms, including a multitude of metabolic and regulatory pathways for its survival. Due to continuing development and improvement in the proteomics technology, difference-gel electrophoresis (DIGE), two-dimensional gel electrophoresis (2D-GE) and liquid chromatography linked to mass spectrometry (LCMS) are now able to detect multiple proteins from complex biological samples (blood, tissues) with higher sensitivity and specificity. Further development of mass spectrometer combined with a technique to remove interfering proteins, such as immunodepleting, will be able to detect even lower concentrations in the samples. Besides, the protein microarray system (chips) is developed as a matrix-support surface to facilitate the binding of selected proteins in matrices of mass spectrometry. These protein chips were developed to bind specific proteins found from the biological samples and identify them very quickly [1]. Proteomics, also known as "complementary to genomics", has been explored enormously for the development of new drug targets. The majority of the currently available drugs target specific proteins of infectious agents than nucleic acid.

The recent development in gel electrophoresis and mass spectrometry has significantly facilitated separation, purification, identification phosphorylation analysis of proteins [2].The transcript detection by mRNA profiling does not reflect the all regulatory process clearly due to post-translational processes altered the amount of active protein insight the cell. The problem could be solved effectively by proteome analysis of the cell. Proteome analysis not only identifies insight changes but also identify post-translational modifications, protein-protein interactions, distribution of cellular and subcellular proteins, and progressive pattern of gene expression also. The utility of differential and/or functional analysis of protein profiling is to acquire hidden cellular information at the atomic level, which will enhance our understanding of the cellular pathways and their inter-relationship with cells/tissue. Through proteomics, it had already been uncovered and validated various potential drug targets against numerous diseases such as TB, HIV, cancer, *etc.*, some of them are available commercially. The current era of proteomics in TB is now beginning to inspect how proteomic technology can help the clinician in biomedical science.

By combining, the advancement of proteomic tools available to the researchers, the entire proteome can now begin to be unraveled the molecular basis of infectious disease; identify the novel biomarkers sets and potential drug targets also. The power of the proteomic approach has been most effectively used to date in the fields of cancer and TB research [2 - 9]. Celis and co-workers [10] have utilized 2-dimensional gel electrophoresis and MS analysis in cancer research to find differentially expressed protein between diseased and healthy tissue including normal urothelium *vs* squamous cell carcinoma (SCCs) that has defined some of the stages involved in the squamous differentiation of bladder transitional epithelium. High through output mass spectrometric analysis of human plasma/serum proteomics is emerging as a potent technique for the identification of different protein profiles in cancer patients. Utilizing a proteomic approach, Nishio *et al.* [11] identified noticeable differences in the phosphorylation status of specific nuclear proteins between drug-resistant and sensitive cell lines. Proteomic analysis was successfully applied to the development of erythromycin resistance in *Streptococcus pneumoniae* (S. pneumoniae) that revealed a significantly increased amount of Glyceraldehyde 3-phosphate dehydrogenase (GAPDH) in the M phenotype of drug-resistant bacterium. It was hypothesized that the GAPDH may provide reducing equivalents for the active efflux mechanism or may even be directly involved in the membrane transport mechanism, which regulates the erythromycin resistance in the M phenotype [12 - 14]. A new function of PstS, a subunit of the phosphate ABC transporter in *S. pneumoniae* penicillin-resistant strains, was revealed by proteomic analysis [15]. Xu and his colleagues [16] reported three antibiotic-resistant proteins of TolC, OmpC, and YhiU, together with the antibiotic resistance-related proteins of FimD (precursor), LamB, Tsx, YfiO, OmpW, and NlpB, which responded to tetracycline and ampicillin resistance in *E. coli* K-12 through proteomic technology.

Proteomics approaches have also been applied to the numerous communicable and non-communicable human diseases such as TB, HIV, Malaria, Leishmaniasis, Toxoplasmosis, cancer, diabetics, *etc.* for development of diagnostic/therapeutic regimens. Numerous potential drug targets were identified by performing comprehensive 2D-gel electrophoresis using cellular and membrane fractions of *P. aeruginosa* clinical isolates [17 - 19]. A similar pattern was reported in *Plasmodium falciparum* also [20, 21].

Tuberculosis (TB) affects millions of people worldwide, caused by *M. tuberculosis*. The appearance of drug-resistant strain of *M. tuberculosis*, including MDR and extensively drug resistant (XDR) strains are a major hurdle to combat for their clearance worldwide. However, the altered protein expression profile of these strains could help the scientific community toward the development of vaccines, serodiagnostic tests [2, 8, 9, 22 - 25], and choice of new drug targets also. The development of drug resistance and the mechanism of action of the drug can be studied through proteomic analysis only. A recent study performed by Agarnoff*et. al.* [26] reported serum proteomic profiles of patients living with *M. tuberculosis* and healthy subjects. They identified a few diagnostic biomarkers using SELDI-TOF-MS with 90% and 94% diagnostic accuracy.

This chapter highlights the prospects of the advances in the proteomics study that could be beneficial to mycobacterial research. It will also provide information to explore the possibility of protein-based biomarkers in the development of new diagnostic and therapeutic for TB.

Advancement of Mycobacterial Research through Proteomics

It is being expected that researchers must compare linear comparison of proteome profile and antigen recognition patterns among or between drug-resistant strains, which would give some clues; 1) why some strains are more prevalent than others? 2) why some are more prone to develop resistance?. The effect observes at the proteomic level will able to suggest conceivable post-translational control mechanisms and reiterates the needs in a more dynamic view of cellular response to drug-related perturbations. Various comparative proteome analysis of *M. tuberculosis* and *M. bovis* BCG strains [27] had revealed the differential expression of 30 conserved hypothetical and six unknown proteins [6, 8, 22, 23, 28 - 30].

Comparative proteomic analysis of different strains of *Mycobacterium* spp. was used as a first step towards the identification and post-genomic characterization of pathologically important strains. The differences in protein volume amongst attenuated and virulent *M. tuberculosis* strains may provide better information toward designing new vaccine candidates or therapeutic agents for TB. Mycobacterial proteins elucidate the compartmentalization of functional networks [31] and reveal the unique proteins in the cytosol, membrane, and cell wall. A total set of differentially expressed proteins will change the way of TB research exclusively and also permit direct analysis of the various expressed gene(s) under strict conditions [2, 8, 32, 33]. Another study performed by Singh *et al.* [2] using comparative proteomic analysis of *M. tuberculosis* reveals the role of various proteins such as wag31, garA, Rv3028c, Rv2970c in the development of INH resistance in *M. tuberculosis.*

PROTEOMICS AND PHYSIOLOGY OF *M. TUBERCULOSIS*

Identification of New Enzymes

The study of Mattow *et al.* [34] compared the cellular proteins profile of each two virulent strains of *M. tuberculosis* and attenuated vaccine strains of *M. bovis* BCG through 2D-GE and identified using mass spectroscopy. The haloalkane dehydrogenase DmbA enzyme was found missing in both attenuated strains (Copenhagen and Chicago), but it was observed in two virulent strains of *M. tuberculosis* ($H_{37}Rv$ and Erdman). It was concluded by the author that lack of the DmbA protein in attenuated strain could be due to low/insignificant gene expression or by the activity of lack repressor. The L-alanine dehydrogenase (Ald, Rv2780) was the first antigen reported to produce by *M. tuberculosis* but absent in *M. bovis* BCG [27, 35]. This suggests excellent diagnostic markers for TB. However, its presence in both virulent as well as avirulent strains of *Mycobacterium.*

New Substrates

Various physiological substrates have been identified through proteomic study and investigated their possible role in various essential gene regulation mechanisms. Singh *et al.* [2], predicted the possible role of GarA (Rv1827) and wag31 (Rv2145c) in drug resistance, especially in MDR-TB cases. They observed that these proteins could serve as physiological substrates for protein kinases G (pknG) and pknB. The FHA domains of Rv0020c and Rv1747 are also predicted as *in vitro* substrate of Ser/Thr kinases *i.e.* pknB, pknD, pknE and pknF [36]. A quantitative proteomics and phosphorylation study identified 265 novel phosphorylation sides derived from 257 proteins in *M. tuberculosis* [37].

New Metabolic Pathways

The genome wide analysis confirms that Ser, Thr and Tyr protein kinases/ phosphatases have been found commonly in the bacterial proteome and can play an important role in cell signaling. The *M. tuberculosis* genome contains genes encoding one phospho-Ser/Thr phosphatase (*PstP*), two phospho-Tyr phosphatases (*ptpA, ptpB*) and 11 Ser/Thr protein kinases (*pknA* to *pknL*), excluding for pknG and pknK, which are soluble. Other residual Ser/Thr protein kinase (STPKs) are predicted as transmembrane "receptor-like" proteins [38]. The pknB is controlled by dephosphorylation and autophosphorylation by the Ser/Thr protein phosphatase pstP [39] which is a major barrier in our understanding of prokaryotes biology of protein kinase. Villarino *et al.* [40] performed proteomic analysis of *M. tuberculosis* and identified GarA, a fork head-associated (FHA) domain-containing protein, which may be served as a putative physiological substrate of pknB.

PROTEOMICS IN DEVELOPING NOVEL BIOMARKERS

Using proteomic techniques, numerous protein-based biomarkers have been identified, which was explored extensively for diagnostic purposes (Table **1**) [2, 25]. However, the majority of these antigens had been banned by the World Health Organization (WHO) [41] due to poor diagnostic value. Among them, few highly antigenic proteins *i.e.* Rv3369 and Rv3374 that are being secreted *in-vitro* by *M. tuberculosis*. The meta-analysis of these proteins had shown poor sensitivity of 60%-73% but better specificity of 96%-97% [4]. Besides the poor sensitivity, these antigens were explored in a rapid kit-based serum screening test but failed due to heterogeneous host immune response against these antigens [42]. Proteomic analysis of culture filtrate proteins reveals the novel proteins are useful for diagnosis and used as a drug target [43].

Table 1. Newly identified biomarkers studied for the serodiagnosis of *M. tuberculosis*.

S. No	Protein/Antigen	Technique	Sensitivity (95% CI)	References
1	Mtb81, MPT32, MPT-64, Ag85A, Ag85C, Ag85, HspX,	ELISA	81	[44]
2	PstS1, Rv0831c, FbpA, EspB, bfrB, HspX and ssb	LIPS	73.5	[45, 46]
3	Rv0054, Rv0831c, Rv2031c, Rv0222, Rv0948c, Rv2853, Rv3405c, Rv3544c	HD-NAPPA	80%	[47]
4	Mtb11, Mtb8, Mtb48	ELISA,	54 -85	[48]
5	Fusion protein (CFP-10, ESat-6, PPE68)	ELISA	73.3%	[49]
6	ESAT-6, CFP-10, ESPC, 14KD/38KD	ELISA	69.4- 90.2	[50]
7	CMX fusion protein (Ag85C, MPT51 and HspX)	ELISA	80.1%	[51]
8	Rv3871, Rv3874, Rv3875, Rv3876, and Rv3879	ELISA	32.21 - 83.56	[52]

S. No	Protein/Antigen	Technique	Sensitivity (95% CI)	References
9	MTB-48kDa, 8kDa, 38kDa, LAM, MPT-64, 16kDa	Rapid test	83.9% to 48.4%	[53]
10	(P38 or PstS1), HspX, Ag85b, MPT32, CFP10, ESAT6, Ag85a, GroES, Rv3507, Rv1926c, Rv2878c, (CFP10-ESAT) fusion, Rv1099, Rv3619, Rv1677, Rv2220, Rv2032, Rv1984c Rv3873, Rv0054, Rv3841, MPT64, Ag85c, Rv1566c, Rv2875, Rv1009, Rv0831c.	Multiplex bead Luminex based assay	88% to 95%	[54]
11	Rv0066c, Rv1310, Rv3375, Rv1415, Rv0567, Rv1886c, Rv3803c, Rv3804c, Rv2031c, Rv1038c, Rv2809, Rv1911c	Proteomics	42% - 76%	[55]
12	Rv3881c, Rv0934, HspX, MPT32, Rv3804c, Ag85a, Rv1886c, Ag85b, Rv0129c Ag85c, ESAT6, CFP10, Rv3841, Rv3418c, MPT70, CFP21, MPT64, Rv0054, CFP10-ESAT fusion, Rv3873, Rv3619, Rv2220, Rv0831c, Rv1009, Rv1099, Rv2032, Rv1926c, Rv2878c, Rv1677, Rv1566c, Rv3507.	In-house multiplex microbead assay	>90%,	[56]

PROTEOMICS AND VACCINE DISCOVERY OF *MYCOBACTERIUM TUBERCULOSIS*

The terms "vaccine" and "vaccinology" came into use soon after Edward Jenner discovered the smallpox vaccine. Its the process of protecting susceptible individuals from the diseases by administration of a living or modified agent (*e.g.*, oral polio vaccine), a suspension of killed organisms (as in pertussis), or an inactivated toxin (as in tetanus) [57]. Vaccines have been one of the main achievements of modern medicine/remedy against deadly infectious diseases such as TB, Hepatitis, polio, *etc.* The millions of deaths were protected using vaccines, suffering from deadly diseases *i.e.* pneumonia, diarrhea, whooping cough, measles, polio, *etc.* Such type of victory not only builds a long history of research and innovation but also boosts the confidence of the scientific community and forced to produce new product discoveries and deliver science figure out ways to reach universal vaccine coverage.

To promote successful and lasting management of TB epidemic, an effective vaccine candidate is required urgently [58]. Although, WHO endorses a singular dose of Bacillus Calmette-Guerin (BCG), a vaccine used in practice for the last 100 years, revaccination with BCG has been standardized in most of the countries. However, it does not offer complete protection against *M. tuberculosis* [58]. The variable efficiency of BCG vaccine indicates that an effective TB vaccine candidate may be possible in the near future. The three different approaches could be accepted for the development of an effective vaccine against TB:

1. To develop an improved version of the BCG vaccine with higher efficacy, which must be safe and long-acting recombinant BCG strains or attenuated *M. tuberculosis*.
2. Prime-boost strategy, typically viral vectored and/or protein adjuvant, would be given to boost our immune system.
3. To develop a therapeutic/passive vaccine (immunotherapeutic), which could reduce the timeline of TB therapy [59].

Table 2. New *M. tuberculosis* vaccine candidates [48].

	Pre-clinical	Phase 1	Phase 2a	Phase 2b	Phase 3
Infants & neonates	BCG-ZMP1 University Zurich, TBVI		MTBVAC Biofabri, Univ. Zaragoza, TBVI		VPM1002 SII, Max Planck, VPM, TBVI
Adolescent & Adults	H107 SSI, TBVI	Ad5 Ag85A McMaster, CanSino	MTBVAC Biofabri, TBVI Zaragoza	M72 + ASO1 GSK, IAVI	VPM1002 SII, Max Planck, VPM
	BCG, ChadOx/MVA PPE15-85A Univ. of Oxford, TBVI	ChadOx1.85A MVA 85A Aerosol / IM Univ. of Oxford, TBVI	TB/Flu04L RIBSP	DAR-901 Dartmouth University, Aeras	MIP Cadila Pharma
	CMV-6Ag Aeras, Vir Biotech, OHSU	GamTBVacMoH Russia	BCG IAVI Revaccination	H56:IC31 SSI, Valneva, Aeras	*M. Vaccae* Anhui ZhifeiLongcom
	CysVac2/Ad University Sydney, TBVI		ID93/GLA-SE IDRI/WT		
	BCG-ZMP1 University Zurich, TBVI				
Therapeutic	MVA Multiphasic vac. Transgene, TBVI	ID93/GLA-SE IDRI / WT	RUTI Archivel Pharma		VPM1002 SII, Max Planck, VPM
	H107 SSI	H56:IC31 SSI, Valneva	TB/Flu04L RIBSP		MIP Cadila Pharma

Since the 1990s, progress has been made toward the development of newer techniques for genetic manipulation of the genome sequence of *M. tuberculosis* [38]. Several effective TB vaccine candidates have been identified, characterized and explored successfully. These candidates are under the trial stage. The detailed information, trial stages are mentioned in Table **2**. The treatment has still delayed when compared to the available resources and research efforts put into other diseases. Progression/trials of vaccine designs rely closely on the outcomes of animal models (murine, bovine and non-primate species). Appropriate animal models are limited because of difficulty to find TB in human species to test on a large scale [60]. The development of potential vaccine candidates is an active area of research and development toward the elimination of *M. tuberculosis* infection and also tackles the spread of drug-resistant strains of *M. tuberculosis*. The relationships between vaccines and its immunogenic response in protection from *M. tuberculosis* infection would significantly facilitate vaccine development. A potent TB vaccine candidate would able to save millions of lives and is highly important toward reaching the Millennium Development Goal of eliminating TB globally by 2050, if developed [58].

PATHWAY OF TB VACCINE DEVELOPMENT

The TB vaccine pipeline requires global and wide-range coordination of efforts with well-defined clinical trial stages of its development and technically defined criteria for the development of specific vaccine candidates. To overcome this problem, TB vaccine development pathway is now established that provides a pre-defined structural path and gating norms for TB vaccine candidates development, validation and evaluation [61]. In tool, the different functions and skills required for the candidate (TB vaccine) and its next stage of development are clearly defined [62] (Fig. **1**).

PROTEOMICS AND HOST-PATHOGEN INTERACTION

Host Cells and Environments for *M. tuberculosis*

TB is a chronic infection developed by the inhalation of infected particles from a patient with active TB. A small drop can spread the alveolar area, which contains almost three bacilli per particle drop. However, it has not known about the minimal infectious dose of TB bacilli for humans [63]. Distinct numbers of bacilli may be required to establish bacilli infection in the human population. It depends on the genetic background of the populations, provinces, dissimilar *M. tuberculosis* strains, virulence, bacilli load, and the host immune response. Therefore, most persons infected with *M. tuberculosis* build up an immune response that is adequate to prevent disease [8]. The explanation to *M. tuberculosis* and the *M. tuberculosis* directed host response shows different clinical outcomes: i) asymptomatic: during infection-sterilization or latent TB infection; ii) symptomatic phase during the manifestation of active infection [63, 64]. Most of the individuals are capable to generate an effective cell-mediated immune response for the control of the infectious stage, whereas approximately 10% of exposed individuals do not generate significant immune response during their lifetime and may develop active disease either as primary infection or reactivation. The other 90% of peoples infected with TB, does not convert into active disease and will never develop an active or transmittable disease. However, They remain latently infected without the entire abolition of *M. tuberculosis* [65, 66]. About 10 percent of immune-competent people infected with *M. tuberculosis*, develop the disease within two years of exposure and having a 10% lifetime risk of developing TB [67].

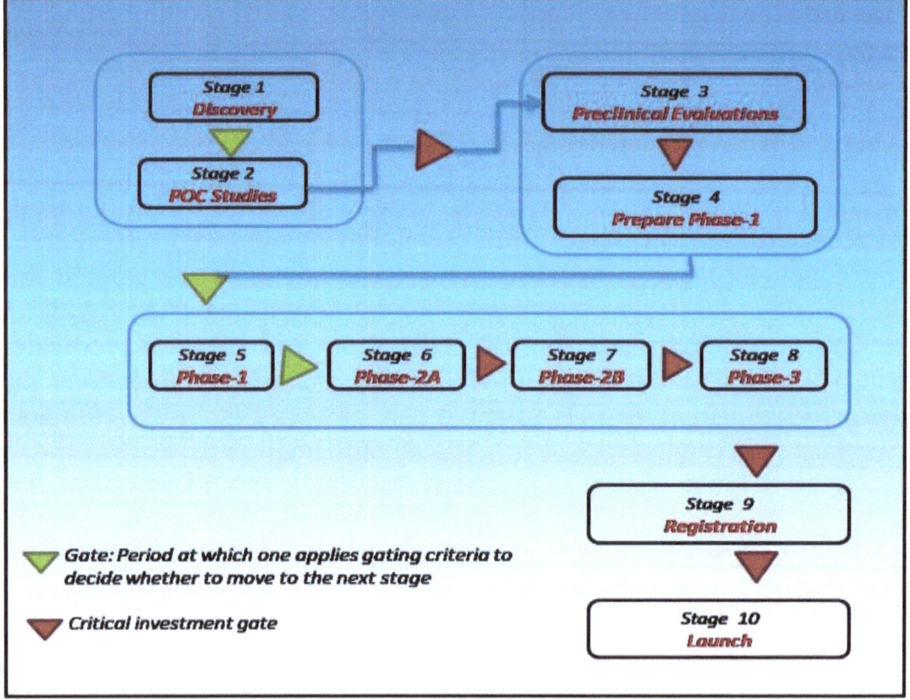

Fig. (1). Pathway of vaccine development.

Once the airborne droplets containing bacilli have been inhaled by the person, bacilli are lodged in the lung parenchyma where they come in contact with the components of the host immune system *i.e.*, neutrophils, macrophages and T - lymphocytes. The *M. tuberculosis* is not only able to modulate the intra-phagosomal activity of macrophages by blocking phagosome maturation but has also evolved specific host immune directed strategies to successfully prevent own clearance by immune cells and to develop long-term survival into the host. The identification and quantitative analysis of complex protein mixtures that were produced during the onset of the infection insight the host are the major challenges due to limited availability and purity of pathogen-containing sections and the irregular ratio of proteins when comparing bacterial and host proteins during the infection.

The alveolar macrophages are the first line of defense against *M. tuberculosis*. Bacilli engulfed by alveolar macrophages and replicate insight into the alveolar macrophage [65, 66, 68]. Since *M. tuberculosis* is an intracellular bacterium that resides inside the macrophages, A major histocompatibility complex (MHC) is needed to present specific *M. tuberculosis* antigens for the activation of the effectors T lymphocytes (CD4+ and CD8+ T-cells) and create an inflammatory response by secreting various cytokines and chemokines [63, 69]. The secreted chemokines (CCL2, CCL3, CCL5, CXCL8, CXCL10) and proinflammatory cytokines (TNF-α, IFN-γ, IL-6, IL-12, IL-17, and IL23) together induce and recruit the immune cells toward the site of infection including T cells and play a major role towards the development of stable localized primary infection (granuloma). The granuloma inhibits the proliferation of mycobacteria and preventing its spread to another organ. The host utilized several cellular and immunological mechanisms to regulate the infection rate, which has been completed with a broad range of *M. tuberculosis* evasion and virulence strategies. If the host successful, suppression of mycobacterial growth occurs, which is called latent infection. If the host immune system failed, *M. tuberculosis* may be reactivated, damage the nearby lung bronchi and infect the other areas of the lungs. Subsequently, bacilli may be entered in to the blood stream and infect other host organs [69].

PROTEOMIC ANALYSIS OF *MYCOBACTERIUM TUBERCULOSIS*-INFECTED CELLS

The intracellular activity of mycobacteria protects from the cellular and humoral response of the host. To overcome host cell defense mechanisms, *M. tuberculosis* modifies the normal passage *via* the endocytic pathway. Moreover, bacilli also capable to induce vacuole rupture to absorb nutrients present insight the cellular cytosol of the host and to escape host defense pathways that support the growth of *M. tuberculosis*. Various proteome based studies of mycobacteria-infected cells were reported to identify host cell factors required for mycobacterial colonization insight the host [70]. Over the past years, a few Liquid chromatography-mass spectrometry (LCMS) studies addressed the relationship between infection settings by the *M. tuberculosis* (both virulent and avirulent strain) which are summarized in Table **3**. Although, proteome analysis of infected host cells/ tissue provides insight information of host defense mechanisms affected by *M. tuberculosis* during the colonization. However, it does not provide detailed information focused on the occupied niche, mycobacteria-infected vacuoles and how its structure and molecular function is misused by *M. tuberculosis* that allows bacilli to grow (replication) insight the host [71].

Table 3. Proteomic analyses of infected host cells with mycobacterial infection performed (A) on total cellular extracts, (B) on isolated cell organelles of infected cells, and (C) specifically on mycobacteria-containing vacuoles (MCVs) and bead-containing phagosomes.

Type of Sample	Experimental Design	Peptide Labeling	Proteins Affected by Infection	Study Findings	Reference
(A) Mass spectrometry (MS)					
Cellular extract	J774.A1 treated by *M. tuberculosis* (lipidome)	iTRAQ/ SILAC	166	oxidation and reduction, signal transduction, vesicle transport, metabolism, *etc.*	[62, 72]
Lung tissue	Tissue infected with *M. tuberculosis* (Granuloma)	No Label	6 peptides of *M. tuberculosis*	Novel *M. tuberculosis* antigens identification, antigen-specific IFN-γ secretion,	[73]
Cellular extract	THP-1 infected by *M. tuberculosis* $H_{37}Rv$ (MOI 1; 1–5 days)	No Label	283	Antimicrobial and inflammatory responses, DNA replication, *etc.*	[74]
Cellular extract	THP-1 infected by BCG/*M. bovis*/*M. tuberculosis* $H_{37}Rv$ (MOI 10; 24 h)	iTRAQ	61	Phagosome maturation and TNF signaling	[29]
(B) MS data of cell organelle proteins					
E.R.	THP-1 infected by $H_{37}Rv$/$H_{37}Ra$	SILAC	133	Cytosolic Ca^{2+} levels, apoptosis, and cholesterol homeostasis	[75]
Exosomes	THP-1 infected with H37Rv	Biotinylated	41	Exosome content analysis	[76]
Secretome	THP-1 infected with H37Rv/H37Ra	iTRAQ		Secretome analysis	[77]
Plasma membrane	THP-1 infected by BCG (MOI 5; 4 h)	SILAC	559	Immune interactions and lipid metabolism	[78]

Type of Sample	Experimental Design	Peptide Labeling	Proteins Affected by Infection	Study Findings	Reference
(C) MS data of MCVs and phagosomes					
Phagosome	THP-1 infected by BCG or beads	SYPRO Ruby	447	Membrane trafficking and signal transduction	[79]
Phagosome	RAW264.7 incubated with beads coated with ManLAM/PILAM/LPS	iTRAQ	42	Vesicle trafficking, phagosome maturation	[80]
Phagosome	RAW264.7 incubated with *M. tuberculosis* or trehalose-dimycolate-coated beads for 30 min	Label-free	835	Mitochondria, cytoskeleton, plasma membrane, ER, endosome, and Golgi-associated proteins	[71]

PROTEOMICS AND STRUCTURAL BIOLOGY

The treatment of *M. tuberculosis* typically comprises a cocktail of anti-TB antibiotics. Most of the antibiotics had developed several years ago and are not effective against dormant tubercle and drug-resistant bacilli, which impends to make many of them superseded. The new drug or drug target identification is very important [81]. Both structural biology and proteomics had been played a significant role in the identification of promising drug targets and consenting the understanding of protein function as well as protein-drug interactions at the atomic level. The detection of *M. tuberculosis*-specific protein/antigen structures has been an objective of the scientific community several years ago, who had expected to provide a huge quantity of structural data which would be useful in identification and designing of newer drugs/drug target.

The existing structural data have already been directed to identifying a number of potential new drug targets and have been helpful in assigning the molecular functions to those proteins, whose function was unknown previously. One of the main breakthroughs in mycobacteria research was the whole genome sequencing of the laboratory strain *M. tuberculosis* $H_{37}Rv$ during 1998, which was re-annotated in 2002 [82]. It offered a boundless opportunity to understand the pathogenesis of mycobacterium at the genomic level, identification of important genes for growth, virulence and persistence. It has been found that ~20% of the proteins are directly involved in virulence and detoxification of *M. tuberculosis*. Later, proteomic data were deposited in the Protein Data Bank (PDB). After the genome sequence became freely available to the user, numerous proteins structure was determined and increased dramatically (Fig. **2**). Initially, structural genomics efforts found low or poor protein solubility, which was a major drawback in the determination of molecular structure. subsequently, various methods have been developed, which eased this premature problem, *i.e* development of customized expression approaches for *M. tuberculosis* proteins in *Escherichia coli*, use of *M. smegmatis* as expression strain, and co-expression system *etc* [82, 83].

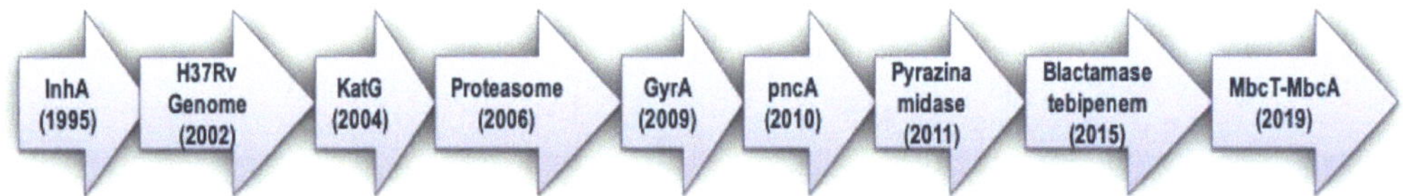

Fig. (2). The progress of *M. tuberculosis* protein structures deposited in the PDB.

STRUCTURE-BASED DRUG DISCOVERY

Interaction between protein structure and ligands or inhibitors can only be determined by structural biology tools. It can provide information about the molecular interactions among them and provide valuable information, particularly when these are potential drug candidates or drugs [84]. It has a long successful history of developing or designing novel drugs such as HIV protease inhibitors agenerase, viracept, aluviran, *etc.*, which had commercialized successfully. Despite these successes, only a few drugs have been developed and validated in this manner. The constant efforts to determine the molecular structure of large numbers of *M. tuberculosis* proteins are based on the probability of developing new drug discovery programs. In the past 20 years, an increasing amount of funds had been invested in structural data on TB proteins worldwide but have yet to lead to a marketable drug due to the absence of fully molecular structure of proteins especially hypothetical/unknown proteins that would be extremely useful in designing of the new drug. To design an effective drug, a high-resolution structure of the protein-ligand complex is a pre-requisite for drug design on structure-based, a large number of *M. tuberculosis* protein structures would be agreeable to such methods [85, 86]. Further information about the *M. tuberculosis* structural genomics data is freely accessible over the following link: http://www.webtb.org/.

Screening for inhibitors against such proteins of *M. tuberculosis* had been performed on several targets. However, in some cases, protein structure is known. This may permit the development of structure-based drug design for *M. tuberculosis* [87, 88]. So far, numerous computational analyses have led to the prediction of probable drug candidates, but this has not yet been verified by actual experiential data. The full potential of structural data to support the new drug discovery against *M. tuberculosis*, therefore, remains to be realized.

Relationship between Proteomics and Genomics

Recent research publications on structural biology and proteomics opened a new way in search of new drug targets. Kruh *et al.* [76] performed a proteomic study on *M. tuberculosis* aerosol challenged guinea pig model. The group had identified more than 500 proteins throughout infection with several classes *i.e.* cell wall and cell processes (Class-3), and intermediate metabolism and respiration (class 7). This indicates the new prospective targets for drug development and structural biology (Table **4**).

Table 4. Functional characterization of structural proteins.

Class	Function	ORFs	Structures	%Structure/ORFs	Folds	Structures/fold
0	Virulence, Detoxification Adaptation	99	20	20.2	15	1.3
1	Lipid Metabolism	233	31	13.3	9	3.4
2	Information pathways	229	25	10.9	9	2.8
3	Cell wall & cell process	708	31	4.4	11	2.8
4	RNAs coding genes					
5	Insertion seq and tag	82	65	44.2		
6	Proline-glutamate (PE) and Proline-Proline-glutamate (PPE) proteins	170	2	1.2	1	2
7	Metabolism, respiration	894	143	1.2	1	2
8	Unknown function	272	0			
9	Regulatory protein	189	32	16.9	11	2.9
10	Conserved hypothetical	1051`	43	4.1	15	2.9

Utility of Proteomics in Drug Discovery

The *M. tuberculosis* genome is encoded by approximately 4000 genes, of which approximately 500 genes encode functional proteins that provide an excellent source for the identification of new targets or development of new/current drugs against *M. tuberculosis*. However, Lamichhane *et al.* [89] reported that TB chemotherapy explored only 10 of these *M. tuberculosis* proteins. Various drug targets and ligands/inhibitors had been explored in the past as strong chemotherapeutic agents, only a few have shown good activity in TB chemotherapy (Table **5**).

Table 5. Summary of these targets, and their current and/or new drug ligands.

	Target	Pathway	Inhibitors	Reference
Targeting actively growing *M. tuberculosis*	GlgE	Maltose metabolism	-	[90, 91]
	Mycolic acid	Mycolic acid metabolism	Dictylamine	[92]
	DprE1/DprE2	Cell wall metabolism	Benzothiazinone, dinitrobenzamides	[93, 94]
	HisG	Histidine Biosynthesis	Nitrobenzothiazole	[95]
	AtpE	ATP synthesis	Diarylquinoline	[96]
	Def	Protein processing	LBK-611	[97]
	MshC	Mycothiol ligase	Dequalinium chloride	[98, 98]
	DprE1	Cell wall synthesis	BTZ042, BTZ169, TBA7371	[93]
	QcrB	Oxidative phosphorylation	Imidazopyridine, Pyrrolo, AX-35	[99]
	GyrA	Protein synthesis	Thiophenes	[100]
	GyrB	Protein synthesis	SPR720	[100, 101]
	LeuRS	Protein synthesis	Gsk656 (oxaborole)	[102, 103]
	Methionine aminopeptidase (MetAP)	Protein processing	2,3-dichliro-1,4-naphthoquinones	[104]
Targeted on Dormant *M. tuberculosis*	Isocitrate lyase	Energy metabolism	-	[79, 105]
	Proteasome complex	Protein processing	Oxathiazol-2-one	[106]
	Transpeptidase	Peptidoglycon metabolism	-	[107]
	DosR	Hypoxia regulation	-`	[108]
	CarD	Stringent response	-	[109, 110]
	MshC	Redox homeostasis	Degualinum	[98]
	Mtr	Redox homeostasis	Benzoisoquinoline-7,12-diones	[102]

Enoyl[acyl-carrier protein] reductase enzyme has been shown as a major drug target due to its major role in the biosynthesis of mycolic acid [111]. Additionally, secretary proteins of *M. tuberculosis* play a fundamental role in host-pathogen interactions and help nutrients during acquisition, direct the host immune response and obstruct the therapeutic intervention [112]. The *M. tuberculosis* secretome consists of multiple proteins that are essential for successful invasion and *in vivo* growth of bacilli during host infection. Therefore, these essential proteins would be the most appropriate drug targets for the development of newer diagnostic tools and/or new drugs [111].

CONCLUSION

This chapter shows the usefulness of proteomic tools together with the computational approach towards finding effective drug targets of *M. tuberculosis* and the development of potent inhibitors from a large dataset of phytochemicals/or synthetic compounds. Further, we explored the utility of *M. tuberculosis* proteome in the identification of new biomarkers drug targets for *M. tuberculosis*. These target/compounds could be used as potential drug targets against *M. tuberculosis* in near future if validated properly.

CONSENT FOR PUBLICATION

Not applicable.

CONFLICT OF INTEREST

The authors declare no conflict of interest, financial or otherwise.

ACKNOWLEDGEMENTS

Declared none.

REFERENCES

[1]   Petricoin E, Wulfkuhle J, Espina V, Liotta LA. Clinical proteomics: revolutionizing disease detection and patient tailoring therapy. J Proteome Res 2004; 3(2): 209-17.
[http://dx.doi.org/10.1021/pr049972m] [PMID: 15113096]

[2]   Singh A, Gopinath K, Sharma P, *et al.* Comparative proteomic analysis of sequential isolates of *Mycobacterium tuberculosis* from a patient with pulmonary tuberculosis turning from drug sensitive to multidrug resistant. Indian J Med Res 2015; 141(1): 27-45.
[http://dx.doi.org/10.4103/0971-5916.154492] [PMID: 25857493]

[3]   Albeldas C, Ganief N, Calder B, *et al.* Global proteome and phosphoproteome dynamics indicate novel mechanisms of vitamin C induced dormancy in *Mycobacterium smegmatis*. J Proteomics 2018; 180: 1-10.
[http://dx.doi.org/10.1016/j.jprot.2017.10.006] [PMID: 29038038]

[4]   Bahk YY, Kim SA, Kim J-S, *et al.* Antigens secreted from *Mycobacterium tuberculosis*: identification by proteomics approach and test for diagnostic marker. Proteomics 2004; 4(11): 3299-307.
[http://dx.doi.org/10.1002/pmic.200400980] [PMID: 15378731]

[5]   Betts JC. Transcriptomics and proteomics: tools for the identification of novel drug targets and vaccine candidates for tuberculosis. IUBMB Life 2002; 53(4-5): 239-42.
[http://dx.doi.org/10.1080/15216540212651] [PMID: 12121002]

[6]   Cho SH, Goodlett D, Franzblau S. ICAT-based comparative proteomic analysis of non-replicating persistent *Mycobacterium tuberculosis*. Tuberculosis (Edinb) 2006; 86(6): 445-60.
[http://dx.doi.org/10.1016/j.tube.2005.10.002] [PMID: 16376151]

[7]   Kumar B, Sharma D, Sharma P, Katoch VM, Venkatesan K, Bisht D. Proteomic analysis of *Mycobacterium tuberculosis* isolates resistant to kanamycin and amikacin. J Proteomics 2013; 94: 68-77.

[http://dx.doi.org/10.1016/j.jprot.2013.08.025] [PMID: 24036035]

[8]   Sharma D, Kumar B, Lata M, *et al.* Comparative proteomic analysis of aminoglycosides resistant and susceptible *Mycobacterium tuberculosis* clinical isolates for exploring potential drug targets. PLoS One 2015; 10(10): e0139414.
[http://dx.doi.org/10.1371/journal.pone.0139414] [PMID: 26436944]

[9]   Lata M, Sharma D, Deo N, Tiwari PK, Bisht D, Venkatesan K. Proteomic analysis of ofloxacin-mono resistant *Mycobacterium tuberculosis* isolates. J Proteomics 2015; 127(Pt A): 114-21.
[http://dx.doi.org/10.1016/j.jprot.2015.07.031] [PMID: 26238929]

[10]   Celis JE, Gromov P. Proteomics in translational cancer research: toward an integrated approach. Cancer Cell 2003; 3(1): 9-15.
[http://dx.doi.org/10.1016/S1535-6108(02)00242-8] [PMID: 12559171]

[11]   Nishio K, Sugimoto Y, Kasahara K, *et al.* Increased phosphorylation of nuclear phosphoproteins in human lung-cancer cells resistant to cis-diamminedichloroplatinum (II). Int J Cancer 1992; 50(3): 438-42.
[http://dx.doi.org/10.1002/ijc.2910500319] [PMID: 1310490]

[12]   Cash P, Argo E, Ford L, Lawrie L, McKenzie H. A proteomic analysis of erythromycin resistance in *Streptococcus pneumoniae*. Electrophoresis 1999; 20(11): 2259-68.
[http://dx.doi.org/10.1002/(SICI)1522-2683(19990801)20:11<2259::AID-ELPS2259>3.0.CO;2-F] [PMID: 10493130]

[13]   Karlsson R, Gonzales-Siles L, Gomila M, *et al.* Proteotyping bacteria: Characterization, differentiation and identification of pneumococcus and other species within the Mitis Group of the genus Streptococcus by tandem mass spectrometry proteomics. PLoS One 2018; 13(12): e0208804.
[http://dx.doi.org/10.1371/journal.pone.0208804] [PMID: 30532202]

[14]   Yang X-Y, Xu J-Y, Wei Q-X, Sun X, He Q-Y. Comparative Proteomics of *Streptococcus pneumoniae* Response to Vancomycin Treatment. OMICS 2017; 21(9): 531-9.
[http://dx.doi.org/10.1089/omi.2017.0098] [PMID: 28934029]

[15]   Soualhine H, Brochu V, Ménard F, *et al.* A proteomic analysis of penicillin resistance in *Streptococcus pneumoniae* reveals a novel role for PstS, a subunit of the phosphate ABC transporter. Mol Microbiol 2005; 58(5): 1430-40.
[http://dx.doi.org/10.1111/j.1365-2958.2005.04914.x] [PMID: 16313627]

[16]   Xu C, Lin X, Ren H, Zhang Y, Wang S, Peng X. Analysis of outer membrane proteome of *Escherichia coli* related to resistance to ampicillin and tetracycline. Proteomics 2006; 6(2): 462-73.
[http://dx.doi.org/10.1002/pmic.200500219] [PMID: 16372265]

[17]   Nouwens AS, Walsh BJ, Cordwell SJ. Application of proteomics to *Pseudomonas aeruginosa*. Adv Biochem Eng Biotechnol 2003; 83: 117-40.
[http://dx.doi.org/10.1007/3-540-36459-5_5] [PMID: 12934928]

[18]   Margalit A, Kavanagh K, Carolan JC. Characterization of the proteomic response of A549 cells following sequential exposure to *Aspergillus fumigatus* and *Pseudomonas aeruginosa*. J Proteome Res 2019.
[http://dx.doi.org/10.1021/acs.jproteome.9b00520] [PMID: 31693381]

[19]   Erdmann J, Thöming JG, Pohl S, Pich A, Lenz C, Häussler S. The core proteome of biofilm-grown clinical *Pseudomonas aeruginosa* isolates. Cells 2019; 8(10): E1129.
[http://dx.doi.org/10.3390/cells8101129] [PMID: 31547513]

[20]   Lasonder E, Ishihama Y, Andersen JS, *et al.* Analysis of the *Plasmodium falciparum* proteome by high-accuracy mass spectrometry. Nature 2002; 419(6906): 537-42.
[http://dx.doi.org/10.1038/nature01111] [PMID: 12368870]

[21]   Rujimongkon K, Mungthin M, Tummatorn J, *et al.* Proteomic analysis of *Plasmodium falciparum* response to isocryptolepine derivative. PLoS One 2019; 14(8): e0220871.

[http://dx.doi.org/10.1371/journal.pone.0220871] [PMID: 31393938]

[22]   Singh A, Gupta AK, Gopinath K, *et al.* Comparative proteomic analysis of sequential isolates of *Mycobacterium tuberculosis* sensitive and resistant Beijing type from a patient with pulmonary tuberculosis. Int J Mycobacteriol 2016; 5 (Suppl. 1): S123-4.
[http://dx.doi.org/10.1016/j.ijmyco.2016.10.028] [PMID: 28043501]

[23]   Pheiffer C, Betts JC, Flynn HR, Lukey PT, van Helden P. Protein expression by a Beijing strain differs from that of another clinical isolate and *Mycobacterium tuberculosis* H37Rv. Microbiology (Reading) 2005; 151(Pt 4): 1139-50.
[http://dx.doi.org/10.1099/mic.0.27518-0] [PMID: 15817781]

[24]   Sharma D, Bisht D. Secretory proteome analysis of streptomycin-resistant *Mycobacterium tuberculosis* clinical isolates. SLAS DISCOVERY: Advancing Life Sciences R&D 2017 Dec; 22(10): 1229-38.
[http://dx.doi.org/10.1177/2472555217698428]

[25]   Singh A, Kumar Gupta A, Gopinath K, Sharma P, Singh S. Evaluation of 5 Novel protein biomarkers for the rapid diagnosis of pulmonary and extra-pulmonary tuberculosis: preliminary results. Sci Rep 2017; 7: 44121.
[http://dx.doi.org/10.1038/srep44121] [PMID: 28337993]

[26]   Agranoff D, Fernandez-Reyes D, Papadopoulos MC, *et al.* Identification of diagnostic markers for tuberculosis by proteomic fingerprinting of serum. Lancet 2006; 368(9540): 1012-21.
[http://dx.doi.org/10.1016/S0140-6736(06)69342-2] [PMID: 16980117]

[27]   Jungblut PR, Schaible UE, Mollenkopf HJ, *et al.* Comparative proteome analysis of *Mycobacterium tuberculosis* and *Mycobacterium bovis* BCG strains: towards functional genomics of microbial pathogens. Mol Microbiol 1999; 33(6): 1103-17.
[http://dx.doi.org/10.1046/j.1365-2958.1999.01549.x] [PMID: 10510226]

[28]   Jhingan GD, Kumari S, Jamwal SV, *et al.* Comparative proteomic analyses of avirulent, virulent, and clinical strains of *Mycobacterium tuberculosis* identify strain-specific patterns. J Biol Chem 2016; 291(27): 14257-73.
[http://dx.doi.org/10.1074/jbc.M115.666123] [PMID: 27151218]

[29]   Li P, Wang R, Dong W, *et al.* Comparative proteomics analysis of human macrophages infected with virulent *Mycobacterium bovis*. Front Cell Infect Microbiol 2017; 7: 65.
[http://dx.doi.org/10.3389/fcimb.2017.00065] [PMID: 28337427]

[30]   Ryoo SW, Park YK, Park S-N, *et al.* Comparative proteomic analysis of virulent Korean *Mycobacterium tuberculosis* K-strain with other mycobacteria strain following infection of U-937 macrophage. J Microbiol 2007; 45(3): 268-71.
[PMID: 17618234]

[31]   Mawuenyega KG, Forst CV, Dobos KM, *et al. Mycobacterium tuberculosis* functional network analysis by global subcellular protein profiling. Mol Biol Cell 2005; 16(1): 396-404.
[http://dx.doi.org/10.1091/mbc.e04-04-0329] [PMID: 15525680]

[32]   Yari S, Tasbiti AH, Ghanei M, *et al.* Proteomic analysis of sensitive and multi drug resistant *Mycobacterium tuberculosis* strains. Microbiology 2016; 85: 350-8.
[http://dx.doi.org/10.1134/S0026261716030164]

[33]   Jiang X, Zhang W, Gao F, Huang Y, Lv C, Wang H. Comparison of the proteome of isoniazid-resistant and -susceptible strains of *Mycobacterium tuberculosis*. Microb Drug Resist 2006; 12(4): 231-8.
[http://dx.doi.org/10.1089/mdr.2006.12.231] [PMID: 17227207]

[34]   Mattow J, Jungblut PR, Schaible UE, *et al.* Identification of proteins from *Mycobacterium tuberculosis* missing in attenuated *Mycobacterium bovis* BCG strains. Electrophoresis 2001; 22(14): 2936-46.
[http://dx.doi.org/10.1002/1522-2683(200108)22:14<2936::AID-ELPS2936>3.0.CO;2-S]   [PMID: 11565788]

[35]   Anderson NL, Anderson NG. A two-dimensional gel database of human plasma proteins. Electrophoresis 1991; 12(11): 883-906.
[http://dx.doi.org/10.1002/elps.1150121108] [PMID: 1794344]

[36]   Grundner C, Gay LM, Alber T. *Mycobacterium tuberculosis* serine/threonine kinases PknB, PknD, PknE, and PknF phosphorylate multiple FHA domains. Protein Sci 2005; 14(7): 1918-21.
[http://dx.doi.org/10.1110/ps.051413405] [PMID: 15987910]

[37]   Verma R, Pinto SM, Patil AH, *et al.* Quantitative proteomic and phosphoproteomic analysis of H37Ra and H37Rv strains of *Mycobacterium tuberculosis.* J Proteome Res 2017; 16(4): 1632-45.
[http://dx.doi.org/10.1021/acs.jproteome.6b00983] [PMID: 28241730]

[38]   Cole ST, Brosch R, Parkhill J, *et al.* Deciphering the biology of *Mycobacterium tuberculosis* from the complete genome sequence. Nature 1998; 393(6685): 537-44.
[http://dx.doi.org/10.1038/31159] [PMID: 9634230]

[39]   Boitel B, Ortiz-Lombardía M, Durán R, *et al.* PknB kinase activity is regulated by phosphorylation in two Thr residues and dephosphorylation by PstP, the cognate phospho-Ser/Thr phosphatase, in *Mycobacterium tuberculosis.* Mol Microbiol 2003; 49(6): 1493-508.
[http://dx.doi.org/10.1046/j.1365-2958.2003.03657.x] [PMID: 12950916]

[40]   Villarino A, Duran R, Wehenkel A, *et al.* Proteomic identification of *M. tuberculosis* protein kinase substrates: PknB recruits GarA, a FHA domain-containing protein, through activation loop-mediated interactions. J Mol Biol 2005; 350(5): 953-63.
[http://dx.doi.org/10.1016/j.jmb.2005.05.049] [PMID: 15978616]

[41]   WHO. Policy statement: commercial serodiagnostic tests for diagnosis of tuberculosis. 2011; 1-26.

[42]   Lyashchenko KP, Greenwald R, Esfandiari J, *et al.* Field application of serodiagnostics to identify elephants with tuberculosis prior to case confirmation by culture. Clin Vaccine Immunol 2012; 19(8): 1269-75.
[http://dx.doi.org/10.1128/CVI.00163-12] [PMID: 22695162]

[43]   Weldingh K, Rosenkrands I, Jacobsen S, Rasmussen PB, Elhay MJ, Andersen P. Two-dimensional electrophoresis for analysis of *Mycobacterium tuberculosis* culture filtrate and purification and characterization of six novel proteins. Infect Immun 1998; 66(8): 3492-500.
[http://dx.doi.org/10.1128/IAI.66.8.3492-3500.1998] [PMID: 9673225]

[44]   Tucci P, González-Sapienza G, Marin M. Pathogen-derived biomarkers for active tuberculosis diagnosis. Front Microbiol 2014; 5: 549.
[http://dx.doi.org/10.3389/fmicb.2014.00549] [PMID: 25368609]

[45]   Burbelo PD, Keller J, Wagner J, *et al.* Serological diagnosis of pulmonary *Mycobacterium tuberculosis* infection by LIPS using a multiple antigen mixture. BMC Microbiol 2015; 15: 205.
[http://dx.doi.org/10.1186/s12866-015-0545-y] [PMID: 26449888]

[46]   Li J-L, Huang X-Y, Chen H-B, *et al.* Simultaneous detection of IgG and IgM antibodies against a recombinant polyprotein PstS1-LEP for tuberculosis diagnosis. Infect Dis (Lond) 2015; 47(9): 643-9.
[http://dx.doi.org/10.3109/23744235.2015.1043941] [PMID: 25958814]

[47]   Song L, Wallstrom G, Yu X, *et al.* Identification of antibody targets for tuberculosis serology using high-density nucleic acid programmable protein arrays. Mol Cell Proteomics 2017; 16(4) (Suppl. 1): S277-89.
[http://dx.doi.org/10.1074/mcp.M116.065953] [PMID: 28223349]

[48]   She RC, Litwin CM. Performance of a tuberculosis serologic assay in various patient populations. Am J Clin Pathol 2015; 144(2): 240-6.
[http://dx.doi.org/10.1309/AJCP22DBRYZQGRBI] [PMID: 26185308]

[49]   Xu J-N, Chen J-P, Chen D-L. Serodiagnosis efficacy and immunogenicity of the fusion protein of *Mycobacterium tuberculosis* composed of the 10-kilodalton culture filtrate protein, ESAT-6, and the extracellular domain fragment of PPE68. Clin Vaccine Immunol 2012; 19(4): 536-44.

[http://dx.doi.org/10.1128/CVI.05708-11] [PMID: 22357648]

[50]   Zhang C, Song X, Zhao Y, *et al.* *Mycobacterium tuberculosis* secreted proteins as potential biomarkers for the diagnosis of active tuberculosis and latent tuberculosis infection. J Clin Lab Anal 2014.
[http://dx.doi.org/10.1002/jcla.21782] [PMID: 25131423]

[51]   Zagmignan A, Costa ACD, Viana JL, *et al.* Identification of specific antibodies against the Ag85C-MPT51-HspX fusion protein (CMX) for serological screening of tuberculosis in endemic area. Expert Rev Clin Immunol 2017; 13(8): 837-43.
[http://dx.doi.org/10.1080/1744666X.2017.1345626] [PMID: 28633546]

[52]   Liu Z, Qie S, Li L, *et al.* Identification of novel rd1 antigens and their combinations for diagnosis of sputum smear-/culture+ TB patients. BioMed Res Int 2016; 2016: 7486425.
[http://dx.doi.org/10.1155/2016/7486425] [PMID: 26885516]

[53]   Bai X-J, Yang Y-R, Liang J-Q, *et al.* Diagnostic performance and problem analysis of commercial tuberculosis antibody detection kits in China. Mil Med Res 2018; 5(1): 10.
[http://dx.doi.org/10.1186/s40779-018-0157-6] [PMID: 29562934]

[54]   Khaliq A, Ravindran R, Hussainy SF, *et al.* Field evaluation of a blood based test for active tuberculosis in endemic settings. PLoS One 2017; 12(4): e0173359.
[http://dx.doi.org/10.1371/journal.pone.0173359] [PMID: 28380055]

[55]   Kumar G, Shankar H, Sharma D, *et al.* Proteomics of culture filtrate of prevalent *Mycobacterium tuberculosis* strains: 2D-PAGE map and MALDI-TOF/MS analysis. Slas Discovery: Advancing Life Sciences R&D 2017 Oct; 22(9): 1142-9.

[56]   Shete PB, Ravindran R, Chang E, *et al.* Evaluation of antibody responses to panels of *M. tuberculosis* antigens as a screening tool for active tuberculosis in Uganda. PLoS One 2017; 12(8): e0180122.
[http://dx.doi.org/10.1371/journal.pone.0180122] [PMID: 28767658]

[57]   Lahariya C. Vaccine epidemiology: A review. J Family Med Prim Care 2016; 5(1): 7-15.
[http://dx.doi.org/10.4103/2249-4863.184616] [PMID: 27453836]

[58]   WHO. Global tuberculosis report. 2019.

[59]   Montagnani C, Chiappini E, Galli L, de Martino M. Vaccine against tuberculosis: what's new? BMC Infect Dis 2014; 14 (Suppl. 1): S2.
[http://dx.doi.org/10.1186/1471-2334-14-S1-S2] [PMID: 24564340]

[60]   McShane H, Williams A. A review of preclinical animal models utilised for TB vaccine evaluation in the context of recent human efficacy data. Tuberculosis (Edinb) 2014; 94(2): 105-10.
[http://dx.doi.org/10.1016/j.tube.2013.11.003] [PMID: 24369986]

[61]   Schrager LK, Harris RC, Vekemans J. Research and development of new tuberculosis vaccines: a review. F1000 Res 2018; 7: 1732.
[http://dx.doi.org/10.12688/f1000research.16521.1] [PMID: 30613395]

[62]   Kaufmann SHE, Dockrell HM, Drager N, *et al.* TBVAC2020: advancing tuberculosis vaccines from discovery to clinical development. Front Immunol 2017; 8: 1203.
[http://dx.doi.org/10.3389/fimmu.2017.01203] [PMID: 29046674]

[63]   Cantini F, Nannini C, Niccoli L, Petrone L, Ippolito G, Goletti D. Risk of tuberculosis reactivation in patients with rheumatoid arthritis, ankylosing spondylitis, and psoriatic arthritis receiving non-ant--TNF-targeted biologics. Mediators Inflamm 2017; 2017: 8909834.
[http://dx.doi.org/10.1155/2017/8909834] [PMID: 28659665]

[64]   Esmail H, Barry CE III, Young DB, Wilkinson RJ. The ongoing challenge of latent tuberculosis. Philos Trans R Soc Lond B Biol Sci 2014; 369(1645): 20130437.
[http://dx.doi.org/10.1098/rstb.2013.0437] [PMID: 24821923]

[65]   Evans JT, Serafino Wani RL, Anderson L, *et al.* A geographically-restricted but prevalent *Mycobacterium tuberculosis* strain identified in the West Midlands Region of the UK between 1995

and 2008. PLoS One 2011; 6(3): e17930.
[http://dx.doi.org/10.1371/journal.pone.0017930] [PMID: 21464965]

[66]   Sagwa E, Mantel-Teeuwisse AK, Ruswa N, *et al.* The burden of adverse events during treatment of drug-resistant tuberculosis in Namibia. South Med Rev 2012; 5(1): 6-13.
[PMID: 23093894]

[67]   Sutherland I. Recent studies in the epidemiology of tuberculosis, based on the risk of being infected with tubercle bacilli. Adv Tuberc Res 1976; 19: 1-63.
[PMID: 823803]

[68]   McDonough KA, Kress Y, Bloom BR. Pathogenesis of tuberculosis: interaction of *Mycobacterium tuberculosis* with macrophages. Infect Immun 1993; 61(7): 2763-73.
[http://dx.doi.org/10.1128/IAI.61.7.2763-2773.1993] [PMID: 8514378]

[69]   Kirby AC, Coles MC, Kaye PM. Alveolar macrophages transport pathogens to lung draining lymph nodes. J Immunol 2009; 183(3): 1983-9.
[http://dx.doi.org/10.4049/jimmunol.0901089] [PMID: 19620319]

[70]   Hoffmann E, Machelart A, Song O-R, Brodin P. Proteomics of *Mycobacterium* Infection: Moving towards a Better Understanding of Pathogen-Driven Immunomodulation. Front Immunol 2018; 9: 86.
[http://dx.doi.org/10.3389/fimmu.2018.00086] [PMID: 29441067]

[71]   Herweg J-A, Hansmeier N, Otto A, *et al.* Purification and proteomics of pathogen-modified vacuoles and membranes. Front Cell Infect Microbiol 2015; 5: 48.
[http://dx.doi.org/10.3389/fcimb.2015.00048] [PMID: 26082896]

[72]   Shui W, Petzold CJ, Redding A, *et al.* Organelle membrane proteomics reveals differential influence of mycobacterial lipoglycans on macrophage phagosome maturation and autophagosome accumulation. J Proteome Res 2011; 10(1): 339-48.
[http://dx.doi.org/10.1021/pr100688h] [PMID: 21105745]

[73]   Yuan C-H, Zhang S, Xiang F, *et al.* Secreted Rv1768 From RD14 of *Mycobacterium tuberculosis* activates macrophages and induces a strong IFN-γ-Releasing of CD4$^+$ t cells. Front Cell Infect Microbiol 2019; 9: 341.
[http://dx.doi.org/10.3389/fcimb.2019.00341] [PMID: 31681622]

[74]   Kaewseekhao B, Naranbhai V, Roytrakul S, *et al.* Comparative proteomics of activated THP-1 cells infected with *Mycobacterium tuberculosis* identifies putative clearance biomarkers for tuberculosis treatment. PLoS One 2015; 10(7): e0134168.
[http://dx.doi.org/10.1371/journal.pone.0134168] [PMID: 26214306]

[75]   Saquib NM, Jamwal S, Midha MK, Verma HN, Manivel V. Quantitative proteomics and lipidomics analysis of endoplasmic reticulum of macrophage infected with *Mycobacterium tuberculosis*. Int J Proteomics 2015; 2015: 270438.
[http://dx.doi.org/10.1155/2015/270438] [PMID: 25785198]

[76]   Diaz G, Wolfe LM, Kruh-Garcia NA, Dobos KM. Changes in the membrane-associated proteins of exosomes released from human macrophages after *Mycobacterium tuberculosis* infection. Sci Rep 2016; 6: 37975.
[http://dx.doi.org/10.1038/srep37975] [PMID: 27897233]

[77]   Kumar A, Jamwal S, Midha MK, *et al.* Dataset generated using hyperplexing and click chemistry to monitor temporal dynamics of newly synthesized macrophage secretome post infection by mycobacterial strains. Data Brief 2016; 9: 349-54.
[http://dx.doi.org/10.1016/j.dib.2016.08.055] [PMID: 27672675]

[78]   Long J, Basu Roy R, Zhang YJ, *et al.* Plasma membrane profiling reveals upregulation of abca1 by infected macrophages leading to restriction of mycobacterial growth. Front Microbiol 2016; 7: 1086.
[http://dx.doi.org/10.3389/fmicb.2016.01086] [PMID: 27462310]

[79]   Lee B-Y, Jethwaney D, Schilling B, Clemens DL, Gibson BW, Horwitz MA. The *Mycobacterium*

bovis bacille Calmette-Guerin phagosome proteome. Mol Cell Proteomics 2010; 9(1): 32-53.
[http://dx.doi.org/10.1074/mcp.M900396-MCP200] [PMID: 19815536]

[80] Shui G, Bendt AK, Jappar IA, *et al.* Mycolic acids as diagnostic markers for tuberculosis case detection in humans and drug efficacy in mice. EMBO Mol Med 2012; 4(1): 27-37.
[http://dx.doi.org/10.1002/emmm.201100185] [PMID: 22147526]

[81] Kerantzas CA, Jacobs WR Jr. Origins of combination therapy for tuberculosis: lessons for future antimicrobial development and application. MBio 2017; 8(2): e01586-16.
[http://dx.doi.org/10.1128/mBio.01586-16] [PMID: 28292983]

[82] Ehebauer MT, Wilmanns M. The progress made in determining the *Mycobacterium tuberculosis* structural proteome. Proteomics 2011; 11(15): 3128-33.
[http://dx.doi.org/10.1002/pmic.201000787] [PMID: 21674801]

[83] Bashiri G, Baker EN. Production of recombinant proteins in *Mycobacterium smegmatis* for structural and functional studies. Protein Sci 2015; 24(1): 1-10.
[http://dx.doi.org/10.1002/pro.2584] [PMID: 25303009]

[84] Ramsay RR, Popovic-Nikolic MR, Nikolic K, Uliassi E, Bolognesi ML. A perspective on multi-target drug discovery and design for complex diseases. Clin Transl Med 2018; 7(1): 3.
[http://dx.doi.org/10.1186/s40169-017-0181-2] [PMID: 29340951]

[85] Lionta E, Spyrou G, Vassilatis DK, Cournia Z. Structure-based virtual screening for drug discovery: principles, applications and recent advances. Curr Top Med Chem 2014; 14(16): 1923-38.
[http://dx.doi.org/10.2174/1568026614666140929124445] [PMID: 25262799]

[86] Ferreira LG, Dos Santos RN, Oliva G, Andricopulo AD. Molecular docking and structure-based drug design strategies. Molecules 2015; 20(7): 13384-421.
[http://dx.doi.org/10.3390/molecules200713384] [PMID: 26205061]

[87] Hecker N, Ahmed J, von Eichborn J, *et al.* SuperTarget goes quantitative: update on drug-target interactions. Nucleic Acids Res 2012; 40(Database issue): D1113-7.
[http://dx.doi.org/10.1093/nar/gkr912] [PMID: 22067455]

[88] Koul A, Arnoult E, Lounis N, Guillemont J, Andries K. The challenge of new drug discovery for tuberculosis. Nature 2011; 469(7331): 483-90.
[http://dx.doi.org/10.1038/nature09657] [PMID: 21270886]

[89] Lamichhane G. *Mycobacterium tuberculosis* response to stress from reactive oxygen and nitrogen species. Front Microbiol 2011; 2: 176.
[http://dx.doi.org/10.3389/fmicb.2011.00176] [PMID: 21904537]

[90] Kalscheuer R, Syson K, Veeraraghavan U, *et al.* Self-poisoning of *Mycobacterium tuberculosis* by targeting GlgE in an alpha-glucan pathway. Nat Chem Biol 2010; 6(5): 376-84.
[http://dx.doi.org/10.1038/nchembio.340] [PMID: 20305657]

[91] Gupta AK, Singh A, Singh S. Glycogenomics of *Mycobacterium tuberculosis*. Mycobact Dis 2014; 4(175): 2161-1068.

[92] Barkan D, Liu Z, Sacchettini JC, Glickman MS. Mycolic acid cyclopropanation is essential for viability, drug resistance, and cell wall integrity of *Mycobacterium tuberculosis*. Chem Biol 2009; 16(5): 499-509.
[http://dx.doi.org/10.1016/j.chembiol.2009.04.001] [PMID: 19477414]

[93] Gawad J, Bonde C. Decaprenyl-phosphoryl-ribose 2′-epimerase (DprE1): challenging target for antitubercular drug discovery. Chem Cent J 2018; 12(1): 72.
[http://dx.doi.org/10.1186/s13065-018-0441-2] [PMID: 29936616]

[94] Manina G, Pasca MR, Buroni S, De Rossi E, Riccardi G. Decaprenylphosphoryl-β-D-ribose 2′-epimerase from *Mycobacterium tuberculosis* is a magic drug target. Curr Med Chem 2010; 17(27): 3099-108.
[http://dx.doi.org/10.2174/092986710791959693] [PMID: 20629622]

[95]   Cho Y, Ioerger TR, Sacchettini JC. Discovery of novel nitrobenzothiazole inhibitors for *Mycobacterium tuberculosis* ATP phosphoribosyl transferase (HisG) through virtual screening. J Med Chem 2008; 51(19): 5984-92.
[http://dx.doi.org/10.1021/jm800328v] [PMID: 18778048]

[96]   Karmakar M, Rodrigues CHM, Holt KE, Dunstan SJ, Denholm J, Ascher DB. Empirical ways to identify novel Bedaquiline resistance mutations in AtpE. PLoS One 2019; 14(5): e0217169.
[http://dx.doi.org/10.1371/journal.pone.0217169] [PMID: 31141524]

[97]   Teo KK, Ounpuu S, Hawken S, *et al.* Tobacco use and risk of myocardial infarction in 52 countries in the INTERHEART study: a case-control study. Lancet 2006; 368(9536): 647-58.
[http://dx.doi.org/10.1016/S0140-6736(06)69249-0] [PMID: 16920470]

[98]   Gutierrez-Lugo M-T, Baker H, Shiloach J, Boshoff H, Bewley CA. Dequalinium, a new inhibitor of *Mycobacterium tuberculosis* mycothiol ligase identified by high-throughput screening. J Biomol Screen 2009; 14(6): 643-52.
[http://dx.doi.org/10.1177/1087057109335743] [PMID: 19525487]

[99]   Foo CS, Lupien A, Kienle M, *et al.* Arylvinylpiperazine amides, a new class of potent inhibitors targeting QcrB of *Mycobacterium tuberculosis*. MBio 2018; 9(5): e01276-18.
[http://dx.doi.org/10.1128/mBio.01276-18] [PMID: 30301850]

[100]   Chopra S, Matsuyama K, Tran T, *et al.* Evaluation of gyrase B as a drug target in *Mycobacterium tuberculosis*. J Antimicrob Chemother 2012; 67(2): 415-21.
[http://dx.doi.org/10.1093/jac/dkr449] [PMID: 22052686]

[101]   Chaudhari K, Surana S, Jain P, Patel HM. *Mycobacterium tuberculosis* (MTB) GyrB inhibitors: An attractive approach for developing novel drugs against TB. Eur J Med Chem 2016; 124: 160-85.
[http://dx.doi.org/10.1016/j.ejmech.2016.08.034] [PMID: 27569197]

[102]   Torfs E, Piller T, Cos P, Cappoen D. Opportunities for Overcoming *Mycobacterium tuberculosis* drug resistance: emerging mycobacterial targets and host-directed therapy. Int J Mol Sci 2019; 20(12): E2868.
[http://dx.doi.org/10.3390/ijms20122868] [PMID: 31212777]

[103]   Yuan T, Sampson NS. hit generation in TB drug discovery: from genome to granuloma. Chem Rev 2018; 118(4): 1887-916.
[http://dx.doi.org/10.1021/acs.chemrev.7b00602] [PMID: 29384369]

[104]   Vanunu M, Schall P, Reingewertz T-H, Chakraborti PK, Grimm B, Barkan D. MapB protein is the essential methionine aminopeptidase in *Mycobacterium tuberculosis*. Cells 2019; 8(5): 393.
[http://dx.doi.org/10.3390/cells8050393] [PMID: 31035386]

[105]   Sharma R, Das O, Damle SG, Sharma AK. Isocitrate lyase: a potential target for anti-tubercular drugs. Recent Pat Inflamm Allergy Drug Discov 2013; 7(2): 114-23.
[http://dx.doi.org/10.2174/1872213X11307020003] [PMID: 23506018]

[106]   Lupoli TJ, Vaubourgeix J, Burns-Huang K, Gold B. Targeting the proteostasis network for mycobacterial drug discovery. ACS Infect Dis 2018; 4(4): 478-98.
[http://dx.doi.org/10.1021/acsinfecdis.7b00231] [PMID: 29465983]

[107]   Gokulan K, Varughese KI. Drug resistance in *Mycobacterium tuberculosis* and targeting the l,d-transpeptidase enzyme. Drug Dev Res 2019; 80(1): 11-8.
[http://dx.doi.org/10.1002/ddr.21455] [PMID: 30312987]

[108]   Ioerger TR, O'Malley T, Liao R, *et al.* Identification of new drug targets and resistance mechanisms in *Mycobacterium tuberculosis*. PLoS One 2013; 8(9): e75245.
[http://dx.doi.org/10.1371/journal.pone.0075245] [PMID: 24086479]

[109]   Priya VGS, Muddapur U, Sonawane K, Mehta M. CarD-a reliable target in *M. tuberculosis*. International Journal of Pharmaceutical Science Invention 2014; 3: 38-46.

[110] Zhu DX, Garner AL, Galburt EA, Stallings CL. CarD contributes to diverse gene expression outcomes throughout the genome of *Mycobacterium tuberculosis*. Proc Natl Acad Sci USA 2019; 116(27): 13573-81.
[http://dx.doi.org/10.1073/pnas.1900176116] [PMID: 31217290]

[111] Yao J, Rock CO. Resistance mechanisms and the future of bacterial enoyl-acyl carrier protein reductase (FabI) antibiotics. Cold Spring Harb Perspect Med 2016; 6(3): a027045.
[http://dx.doi.org/10.1101/cshperspect.a027045] [PMID: 26931811]

[112] Stanley SA, Cox JS. Host-pathogen interactions during *Mycobacterium tuberculosis* infections. Curr Top Microbiol Immunol 2013; 374: 211-41.
[http://dx.doi.org/10.1007/82_2013_332] [PMID: 23881288]

CHAPTER 11

Proteome Based Insights in Drug-Resistant *Mycobacterium tuberculosis*

Apoorva Narain[1], **Surya Kant**[1], **Rikesh Kumar Dubey**[2], **Kanchan Srivastava**[1] and **Anand Kumar Maurya**[3,*]

[1] *Department of Respiratory Medicine, King George Medical University, Lucknow, India*

[2] *Division of Microbiology, CSIR-Central Drug Research Institute, Lucknow, India*

[3] *Department of Microbiology, All India Institute of Medical Sciences, Bhopal, India*

Abstract: *Mycobacterium tuberculosis* (MTB) has been an exceptionally successful human pathogen over the centuries, infecting almost one-third of the global population. An exponential increase in tuberculosis (TB) cases, mainly by the drug-resistant (DR) strains of MTB, has created an urgent need for identifying and developing new antituberculosis drugs acting *via* novel mechanisms. The multi drug-resistant (MDR) TB and extensively drug-resistant (XDR) TB strains, accelerating through drug specific resistance amplification, are resistant to a majority of antibiotics used in the treatment and are challenging to remove from the host's system. Since proteins are the functional beings of the biological arrangement, they make promising drug targets for immunodiagnostics or therapeutics. To identify and characterize such novel proteins, which directly or indirectly regulatedrug resistance in mycobacteria, proteomics approaches could be successfully employed. Serological techniques like immunoassay have higher chances of rendering false positive or false negative results and hence could be rectified by using more sophisticated techniques like mass spectrometry. In the past two to three decades, proteomics-based approach has seen a pivotal rise. The application of proteomics-based approaches has helped to gain insights into MTB and its relevance to clinical science. They have aided in the identification and characterization of novel proteins. To have a better understanding of pathophysiology of MTB, proteome-based science could help simultaneously in the identification of proteins, which can be potential targets. Recent progress in the area of proteomics has opened up the doors to address many previously unanswered questions, with studies on DR-TB being no exception. The Beijing family of MTB forms an interesting candidate for proteomic analysis as it constitutes 13% of the global isolates and has higher chances of acquiring drug resistance. Proteomics can play an important role in the discovery of biomarkers for TB and other diseases. Also, it can aid in the development of effective vaccines as well as simple, rapid, and cost-effective tests for the diagnosis of TB, which are crucial for the management and control of the disease.

* **Corresponding author Anand Kumar Maurya:** Department of Microbiology, All India Institute of Medical Sciences, Bhopal, India; Tel: 0755 2672328; E-mail: anandmaurya1@gmail.com

Divakar Sharma (Ed.)
All rights reserved-© 2020 Bentham Science Publishers

Keywords: Diagnosis, Drug resistance, Host-pathogen interaction, *Mycobacterium, tuberculosis*, Proteomics, Tuberculosis.

INTRODUCTION

Mycobacterium tuberculosis (MTB) is a facultative intracellular pathogen that has the ability to survive inside the macrophages of its host. Though tuberculosis (TB) has been known to have emerged around 70,000 years ago, it still remains a global issue. TB has a mortality of ~70%, if left untreated or improperly treated.

It is the leading cause of death by a single infectious agent and has killed about 1 billion people over two centuries. The impact of the disease is such that it has even slowed down the GDP of countries [1]. Robert Koch, in 1882, used the term tuberkulose to describe his discovery of the bacterium he called Tubercle bacillus. He was conferred with Nobel Prize in physiology/medicine for this landmark discovery in 1905. This discovery, in association with the discovery of tuberculin, the Bacillus-Calmette Guérin (BCG) vaccine, and anti-tuberculosis drugs in the years 1890, 1908, and 1943, respectively, paved a way to combat the deadly disease of tuberculosis. The main causal organism of TB in humans is *Mycobacterium tuberculosis*, though other *M.tuberculosis* complex (MTBC) members (*M.bovis, M.microti and M.africanum*) can also cause tuberculosis. MTB, being an intracellular pathogen, interacts with host's immune response to cause major immunopathological symptoms. To establish successful infection, MTB have adapted several strategies to survive intracellularly, such as inhibition of phagolysosomal fusion, maturation of phagosomes, drug efflux, and targeting mitochondria to disturb the balance of pro-apoptotic and anti-apoptotic factors. All these cellular events are primarily regulated by bacterial proteins. Nearly 40% of the proteins identified in bacteria are uncharacterized in MTB and approximately 50% of the proteins belong to this earlier mentioned category [2]. Proteins have a responsible astonishing range of biological functions and roles, such as structural proteins, enzymes, and transportation. To promote a healthy growth rate and intracellular survival, every protein in the cell interacts with the cellular environment. Every protein is expressed *via* a gene that forms an interface with all other cellular components to promote the growth and survival of a cell. Hypothetical or uncharacterized category of proteins could be unique to bacteria since they normally have no hits in other genomes, such as hypothetical proteins of the PE/PPE family in mycobacteria [3]. The term proteome was first introduced by Mark Wilkins in 1986 and the most noteworthy contribution was made by R. Aebersold *et al.* in the development of proteomics of mycobacteria. In the 20th century, top-down proteomics was primarily employed for MTB. It works on the principle of separating proteins on the basis of their physical and chemical properties, *i.e.*, gel electrophoresis and gel filtration and ultimately identifying them through mass spectrometric (MS) methods. This way, around 3% of the total proteome of MTB (100 mycobacterial proteins) has been identified. Another type of strategy is known as bottom-up proteomics, which works on the principle of isolating total protein from the sample, proteolytically cleaving it into peptides and then analyzing using high performance liquid chromatography, coupled with tandem mass spectrometry (HPLC- MS/MS) [4, 5].

DRUG RESISTANT TUBERCULOSIS

Mycobacterium spp. possesses intrinsic resistance against a varied range of antibiotics, mainly due to the presence of thick cell walls that are rich in mycolic acid and drug efflux pumps. Some antibiotics become inactive after they are cleaved or modified upon penetrating the cell membrane. The absence of horizontal gene transfer in MTB directs towards chromosomal aberrations being responsible for drug-resistant phenotype instead of resistant plasmid or transposons. Mycobacterial cell wall composition and a low number of porins heavily contribute to cell membranes less permeability to compounds. The lipid layer of the cell wall is linked to the peptidoglycan layer *via* arabinogalactan. Small hydrophilic compounds enter the extremely hydrophobic cell wall by water-filled porins [3, 6]. Drugs able to penetrate the thick cell wall of mycobacteria are enzymatically inactivated inside the bacteria. β-lactamases are one such group of enzymes. They target the antibiotics with β- lactam ring by hydrolyzing them. MTB protein, BlaC (β-lactamase), is localized in the periplasmic space either in free form or as a liposome. Clavulanate, a known inhibitor of β- lactamase, irreversibly inhibits the activity of BlaC. MTB β-lactamase has broad-spectrum substrate specificity and the upscaling of drug-resistant cases in TB has opened the doors of discussion for β-lactam antibiotics to be introduced in the drug regimen. Antibiotics are also rendered inactive by chemical modification, acetylation, or methylation [6, 7]. Enhanced intracellular survival protein (Eis) of MTB acetylates second-line TB drugs kanamycin A and capreomycin, thereby inactivating them in the process [6]. Efflux pumps (EPs) are the membrane-associated transporters that pump out a wide range of molecules from the bacteria system, including drugs, into the outside environment. Efflux systems in *M. tuberculosis* are essential for intracellular growth in macrophages and get activated by immune or drug pressures. Upon activation, they decrease the accumulation of drugs, thereby reducing the cytoplasmic concentration to sub-inhibitory levels. Bacterial efflux pumps have been divided into five super-families: resistance nodulation division (RND), multidrug and toxic compound extrusion (MATE) family, ATP-binding cassette (ABC), small multidrug resistance (SMR), and the major facilitator super-family (MFS) [2]. The average transcription and replication rate in mycobacteria slows down in the presence of drugs, which cause the reduced ribosomal proteins' expression and fluoroquinolones (FQ) target *gyr*A and *gyr*B. For mycobacteria to attain *de novo* resistance against drugs, three major factors come into foray, which include (a) the population size, (b) the mutation rate and (c) mutational target size [2].

CLINICALLY RELEVANT MTB PROTEINS

MTB comprises of a wide array of proteins, lipids, and polysaccharides of antigenic nature. The MTB proteins (6kDa, 10kDa, 16kDa, 19kDa, 23kDa, 35kDa, 47kDa and 85kDa) are being used for immune-diagnostics, for example, diagnosis in immunohistochemical assays, skin patch tests, ELISA and PCR based assays. Most routinely used proteins for diagnosis are the early secreted antigenic target (ESAT-6), culture filtrate protein (CFP-10), Mpt-64, and antigen 85 complex (Ag85).

Antigens of 85 Complex

Proteins of the Ag85 complex, *i.e.*, Ag85A, Ag85B, Ag85C, have been widely studied in TB. Schwander *et al.* and Antas *et al.* have both demonstrated a significant rise in IFN-γ production upon stimulation with the proteins of Ag85 complex, specifically Ag85A and Ag85B, in treated patients (49) as compared to healthy controls (19). Additionally, Alvarez-Corales *et al.* also reported an increase in IL-17 production along with increased IFN-γ production in treated patients when exposed to Ag85A and Ag85B [9].

Mpt64

Mpt64 is a protein secreted specifically by MTBC species. Rv1980c gene encodes the Mpt64 at an earlier stage of the bacterial cycle. In tuberculin skin tests, the protein gives a sensitivity of 87.8% and a specificity of 100%. A few variants of BCG vaccines also express Mpt64 protein, making anti-Mpt64 antibody essential for TB diagnosis. Mpt64 is of specific importance because it can differentiate between MTBC and NTM while also performing well in HIV infected patients [10, 11].

ESAT-6 and CFP-10

Mycobacterial ESAT-6, a 6kDa secretory protein, imparts virulence to mycobacteria by playing an important role in breaching cell membrane integrity, evasion from phagolysosomes, immunomodulation, and bacterial dispersion. It is secreted in the cytoplasm of the host post-infection of macrophages. A conserved mycobacterial region, region of difference 1 (RD1), encodes ESAT-6 together with its whole secretory apparatus. It is well known that it interacts with host proteins, such as syntenin-1, TLR-2, metalloprotease, ADAM9, Beta-2-Micro globulin (β2M) [11, 12]. CFP-10, a 10kDa mycobacterial secretory protein, is also encoded by the RD-1 region and has been frequently used in QuantiFERON assays, ELISpot, and flow cytometry. The diagnosis finds its main application in endemic areas for the diagnosis of pleural effusion that has arisen due to TB. CFP-10, in association with ESAT-6, is one of the most immunogenic molecules of culture filtrate that interacts with host proteins, like laminin, leading to cell lysis that helps in the dissemination of MTB in the lungs [12, 13].

PROTEOMIC ANALYSIS IN MYCOBACTERIA

The proteomics approach is one of the most heavily employed methodologies today to elucidate the function of a gene, even though it is much more complicated than genomics. As genomics has revolutionized the concept of the comprehensive analysis of biological processes, similarly, identification of post-transcriptional mechanisms have led to the finding that direct measurement of protein expression is equally essential in examining biological processes. It analyzes the relation between mRNA and protein levels expressed by a particular gene. It also characterizes the functional properties of the genome. Studying prokaryotic proteins is relatively difficult due to variability in molecular size, hydrophobicity, quantity, and hydrophillicity. The quantitative proteomic analysis includes separation of proteins by isoelectric focusing or gel electrophoresis with mass spectrometric identification of selected protein spots. As proteomics has prognostic and diagnostic value, it carries a notable role in drug development. It is more reliable than genomics or transcriptomics since it is subject to more fluctuations and, as proteins are the products of gene expression, they are tightly regulated by mRNA level and host translational control and regulation [2, 7].

Sodium Dodecyl Sulphate-Poly Acrylamide Gel Electrophoresis

Sodium Dodecyl Sulphate Poly Acrylamide Gel Electrophoresis (SDS-PAGE) is a high resolving technique developed by Ulrich K. Laemmli that separates proteins according to their sizes. Different proteins are separated according to the difference in their charge/mass ratio. Sodium dodecyl sulfate (SDS) has the ability to denature and linearise the proteins, which are then separated *via* PAGE according to their molecular weight. SDS-PAGE possesses diagnostic value, especially in the protein profiling of *Mycoplasma bovis* and *Mycoplasma agalactiae* as a routine diagnostic procedure. It is a discontinuous electrophoretic system that works on separating the biomolecule on two different concentrations of polyacrylamide. Discontinuity is dependent on factors, such as electrode buffer, structure of the gel, pH value, and ionic strength of the buffer and the nature of ions in the gel. Glycine, in electrophoresis buffer, has low mobility and a net charge of pH 6.8 in stacking gel. Proteins in the gel are separated by following the principle of isotachophoresis and form stacks as per stacking effect, *i.e.*, in the order of mobility. Protein mobility is dependent on the m/z ratio and not on the size of the molecule [5, 14].

Two Dimensional Gel Electrophoresis

Another technique that utilizes the principle of m/z ratio for protein separation is two-dimensional gel electrophoresis (2D-GE). What differentiates it from SDS-PAGE is that it first separates the proteins as per their pI in the first dimension and then according to their molecular weight in the second dimension. Based on the size of the gel, it can separate ~5,000 different proteins successively. Neidhardt and Van Bogelen studied bacterial physiology using 2D-GE. The technique has been successfully employed for characterizing metabolic pathways, post-translational and mutant proteins. Virulence factors (like exotoxins) released by S.aureus strains, cell wall membrane proteins of Listeria spp, PyrR bacterial regulatory protein, and metabolic system of *B. subtilis* have all been studied using 2D-GE [5, 14]. A study conducted by Kumar *et al.* showed that 2D-GE, along with mass spectrometry, is still one of the best approaches for proteomic analysis [15 - 17]. In the study, through proteomic analysis of MTB isolates resistant to second-line TB drugs (amikacin and kanamycin), they identified the MTB proteins, namely Rv3867, Rv1932, Rv3418c, Rv1876, Rv2031c, Rv0155, Rv0643c, Rv3224, Rv0952, and Rv0440, showing increased expression. The study also highlighted the role of hypothetical proteins and iron metabolism regulating proteins in providing resistance against the second-line drugs. Proteomic analysis of culture filtrates of aminoglycoside resistant clinical isolates of MTB revealed that a large number of proteins were differentially expressed [18, 19]. Recently, a rapid and efficient method was developed to enrich the lipophilic proteins' extraction from MTB for 2D-GE. These methods utilize an optimized extraction buffer, which increases the extraction of proteins and improves the resolution quantitatively as well as qualitatively on the 2D gel by up to two to three folds [20].

Western Blotting

One essential technique to detect and estimate proteins (antigen) is Western blotting. It employs another important technique in proteomics as well, *i.e.*, SDS-PAGE, to separate the proteins. After separation, the proteins are transferred either onto a PVDF membrane or a nitrocellulose membrane (known as the blots). The proteins that are transferred on the blot are then detected by probing it with the antibody specific to the protein (antigen) of interest, like, capsid proteins of PPV (infects Nicotianabenthamiana) and immunological prevalence of HSV-2 in Africa were identified using 2D-PAGE [5].

Enzyme-Linked Immuno Sorbent Assay

In 1971, Engvall and Pearlmann, by making use of alkaline phosphatase, quantified the IgG in the sera obtained from the rabbit. Enzyme Linked Immuno Sorbent Assay (ELISA) is a highly sensitive immunoassay that is frequently used in diagnostics. It is based on the principle of detecting antigen or antibodies fixed on the solid surface by secondary antibodies conjugated with enzymes and measuring the fluctuations observed in the enzyme activities. The variations quantified in the enzyme activity are directly proportional to the concentration of antigen and/or antibody in the sample. ELISA technique has many variants like direct ELISA, indirect ELISA, capture ELISA, competitive ELISA, sandwich ELISA, and digital ELISA. Making use of ELISA, the surface antigens of Mycobacterium avium subspecies paratuberculosis were detected and the diagnosis of paratuberculosis (John's disease) was facilitated [11, 21]. Recently, antibody response, against three proteins, namely ESAT-6, CFP-10 and purified protein derivative (PPD), was evaluated and the study revealed that PPD was a better antigen than ESAT-6 and CFP-10. It established the fact that utilization of a single protein, either ESAT-6 or CFP-10, is not effective for serodiagnosis of active TB. Identifying unknown antigens of PPD other than ESAT-6 and CFP- 10 will be a good strategy for diagnostic purposes [22].

Immunohistochemistry

Immunohistochemistry (IHC) detects the antigens in tissue sections *via* specific antibodies. As the name indicates, it is an amalgamation of 3 major disciplines of science, namely, immunology, histology, and chemistry. It has been more than 50 years since the technique was first introduced. The ability to specifically detect the antigen in tissue provides it an edge over other protein detection methods. This is very important for the study of cell function in normal and pathological tissues. IHC has been applied extensively to characterize and study cell function. The technique makes use of both monoclonal and polyclonal antibodies. Immunohistochemical methods are being broadly used for the detection of pathogens like cytomegalo virus, hepatitis B/C virus, and in cancer diagnosis. The detection is based on the use of antibodies specific to microbial nucleic acids (DNA or RNA). It is also used for identifying the antigens in cytological preparations. In 1940, the idea of IHC was first introduced, when the pneumococci were identified by using specific antibodies tagged with fluorescein dye. At the time, IHC was known as "immunofluorescence assay" and used to identify certain pathogens in both human and veterinary medicine. For diagnosing an immunocompromised patient, IHC is particularly valuable as it can provide an exact confirmation of infection since it works in the principle of specific antigen-antibody binding [23]. The high sensitivity of IHC for nontuberculous mycobacteria (NTM) compared to MTB suggests that a case with high clinical and histological suspicion for MTB should be sent for PCR in spite of getting a negative result for AFB and IHC [24]. Another study showing the importance of IHC over ZN staining revealed that 72% of cases were identified by IHC while only 27% were identified by ZN staining of the selected samples of the 100 suspected TB cases [25].

Mass Spectrometry

The technique mass spectrometry (MS) measures the m/z ratio of proteins that are then used to ascertain the molecular weight of proteins. Firstly, the molecules are converted into gas-phase ions and are then separated in the mass analyzer based on their m/z values. The separation of molecules can be done in two ways- either in the magnetic field or the electric field. Lastly, the m/z value of separated ions and their amount is calculated. Surface-enhanced laser desorption/ionization (SELDI), electrospray ionization (ESI), and matrix-assisted laser desorption ionization (MALDI) are some of the most frequently used ionization techniques in proteomics. PCR and rRNA gene sequencing, when conjugated to electrospray ionization mass spectrometry (ESI-MS), makes the technique a highly precise one. The technique has been successfully used to characterize the medically important filamentous fungi, yeast, and prototheca. The technique could even detect the post-translational modifications in the proteins. Interleukin-12 (IL- 12), human growth hormone (hGH), prostate-specific antigen (PSA), and blood proteins (IBP2, IGF2), the proposed biomarkers for breast cancer, have all been characterized using MS. Imaging MALDI mass spectrometry is used to analyze the drug and metabolite distribution inside the whole body and also to study the whole body tissues [26].

Isotope-coded Affinity Tag Labeling

Isotope-coded affinity tag (ICAT), the first chemical labeling method to make use of biotin tag to label cysteine moieties (*in vitro*), overcomes the drawbacks of 2D-GE, *i.e.*, identifying proteins that do not come in the pI and molecular weight range of 2D-GE. ICAT tags comprise of three functional moieties, a cysteine reactive site which forms a bond with reduced cysteine in protein, a linker containing either 8 H atoms or 8 ^{2}H atoms, and a biotin affinity tag that allows proper isolation of tagged peptide. This quantification method permits measuring spectra between two different cellular states by a difference of 8 Da. The cysteine residues are labeled by ICAT reagents at C12 and C13, trypsin digested (by ion-exchange chromatography). By using avidin, the peptides that contain cysteine are isolated by affinity chromatography and the C-12/C-13 labeled isotopes are estimated by exercising LC-MS/MS technique. One disadvantage this technique carries is that it specifically only targets those cysteine residues which are available for reaction with the isotope-coded affinity tag [27, 28].

Stable Isotope Labeling by/with Amino Acids in Cell Culture

Stable Isotope Labeling by/with Amino acids in Cell (SILAC) involves metabolic labeling of whole cellular culture proteome, which is then quantified *via* MS analysis. Light or heavy forms of amino acids are used to label proteomes from different cell types. The technique is used to study PTMs, secretory proteins in cell culture, cell signaling, and gene expression regulation. For example, 1500 proteins of two different physiological states of *B.subtilis*, *i.e.*, growth during phosphate starvation and succinate starvation, were quantified *via* SILAC as approximately three-fourth of its genes get expressed during exponential phase. Wild-type, recombinant, and immensely purified protein V from a mutant adenovirus were also measured through SILAC. Dynamic SILAC, an improvement over conventional SILAC, has been used to study the overall protein turnover rate and the stability of many different proteins inside the cellular environment from human adenocarcinoma cells [5].

Isobaric Tags for Relative and Absolute Quantitation

Another MS (specifically tandem mass spectrometry) dependent protein quantification labeling technique is Isobaric tags for relative and absolute quantitation (iTRAQ). The protein is labeled with 8-plex and 4-plex isobaric tags for absolute as well as relative quantification. The amine groups present on the side chain and the N-terminus of a protein are labeled, then fractionated through LC, and ultimately estimated *via* mass spectrometry. 181 membrane proteins and 783 cytosolic proteins of *Thermobifida fusca* (a thermophilic bacillus) have been identified through this approach. The identified proteins were involved in all the basic cellular pathways like glycolysis, TCA cycle, central dogma, purine and pyrimidine synthesis, and pentose phosphate pathway. Deciphering the functioning and subsequent molecular processes involved in the development of natural killer (NK) cells has been made possible *via* iTRAQ labeling. Similarly, in a study (using iTRAQ labeling) conducted on mouse liver, regeneration, following partial hepatectomy, divulged a total of 827 proteins [8]. Wenqing Shui *et al.* conducted a study to analyze the expression of macrophage protein in response to MTB infection making use of state-of-the-art quantitative proteomic approaches, chemical isobaric tagging iTRAQ, and metabolic labeling SILAC. The study showed that only 463 proteins of the total 1286 identified proteins were discovered by both isotope and labeling strategies, while the remaining proteins were discovered only by one of the two approaches [29].

X-ray Crystallography

One of the most common and oldest techniques used to understand the 3D structure of proteins is X-ray crystallography. The diffraction patterns correspond to the size of repeating units that model the crystal (highly purified). X-ray crystallography has a wide array of applications like interpreting the enzyme kinetics, SDM, protein-ligand interaction, designing drugs, studying immune complexes, protein-nucleic acid complexes, and viruses. It has been very useful in identifying the structure of the C-terminal fragment of FtsZ and the binding complex of FtsZ-ZipA. Both FtsZ and ZipA regulate cell division in bacteria. A study on the Norwalk virus (causes gastroenteritis in humans) *via* X-ray crystallography has revealed the 180 repeating units of a single protein that forms the viral capsid. The structure of microsomal cytochrome P450 3A4, a protein that could oxidize a large number of substrates (macrolide antibiotics, statins, cyclosporin) inside a cell, was also decoded using X-ray crystallography [30].

Nuclear Magnetic Resonance Spectroscopy

Nuclear magnetic resonance (NMR) spectroscopy is used to study the folding, behavior, and structure of proteins at the molecular level. The structure of the measured samples is identified by interpretive approaches. When used with liquid chromatography, NMR gives better sensitivity and resolution for high-throughput protein profiling. For example, NMR and UHPLC were used to identify the potential biomarkers in esophageal cancer patients. Interaction between iso-1 cytochrome c and CytC peroxidase from yeast and structure of the transmembrane domain of OMP -A from *E. coli* have also been deciphered *via* NMR and associated high throughput proteomics' techniques. Holmes *et al.* used the NMR technique to describe variations amongst metabolic phenotypic from 4,630 participants belonging to 4 different types of the human population [5, 6]. The solution structures of various proteins from MTB have been studied using NMR spectroscopy. The acyl carrier protein AcpM (Rv2244) of MTB was the protein whose structure was determined by solution NMR in 2002 [31]. Acyl carrier proteins (ACPs) are responsible for the transportation of intermediates between type II fatty acid synthases [32]. Interaction studies of various MTB proteins with other molecules have been conducted based on the proteins observed *via* NMR methods, such as to understand the interaction between MTB Rv2050 protein and σ subunits [33], relationship of the STAS domain of Rv1739c with GTP or GDP [34] and the binding of MptpA with phosphate ion (Pi) as a competitive inhibitor [35].

PROSPECTIVE DIRECTION

Proteomics represents the expression of molecules directly influencing the cell phenotype and is hence, of clinical relevance. These techniques are applicable in investigating a wide array of proteins and related functions both *in vitro* and in *in vivo*. It is of special importance since it also highlights molecules that may otherwise be overlooked in hypothesis-driven scenarios. Proteomics is still a relatively new field of science when compared to genomics and has shortcomings of its own mainly due to low reproducibility. Identifying weakly expressed molecules through MS, though, remains problematic; it could be overcome *via* western blotting. Proteomics, as an investigative technique, is on the rise as it displays the association between identified molecules, generates a large set of databases, and contributes in highlighting bioinformatics as a specialist field. This makes it particularly important in studying disease biology and therapeutics. But when assessing biomarkers, the data validation uses high-throughput techniques, a large sample population and further validation by ELISA, western blotting, and immunohistochemistry, making the whole process time-consuming. Recent advancements in proteomics have paved the way for new ventures in the field of tuberculosis, making it relatively easier to tackle complex issues, especially related to host physiology. Despite being new, a large number of studies involving proteomic approaches are being conducted to study the infectious pathogens and their related pathophysiology. Deciphering MTB pathogen at proteomics level could help in a better understanding of the pathogenic mechanisms that allow MTB to immunomodulate the host's cellular processes, their survival, and also in screening/developing prospective drug targets.

CONSENT FOR PUBLICATION

Not applicable.

CONFLICT OF INTEREST

The authors declare no conflict of interest, financial or otherwise.

ACKNOWLEDGEMENTS

We would like to acknowledge the institutes of all the respective authors.

REFERENCES

[1]   WHO global TB report. 2019.

[2]   Meier NR, Jacobsen M, Ottenhoff THM, Ritz N. A systematic review on novel *Mycobacterium tuberculosis* antigens and their discriminatory potential for the diagnosis of latent and active tuberculosis. Front Immunol 2018; 9: 2476.
[http://dx.doi.org/10.3389/fimmu.2018.02476] [PMID: 30473692]

[3]   Bekmurzayeva A, Sypabekova M, Kanayeva D. Tuberculosis diagnosis using immunodominant, secreted antigens of *Mycobacterium tuberculosis*. Tuberculosis (Edinb) 2013; 93(4): 381-8.
[http://dx.doi.org/10.1016/j.tube.2013.03.003] [PMID: 23602700]

[4]   Cole ST, Brosch R, Parkhill J, *et al.* Deciphering the biology of *Mycobacterium tuberculosis* from the complete genome sequence. Nature 1998; 393(6685): 537-44.
[http://dx.doi.org/10.1038/31159] [PMID: 9634230]

[5]   Aslam B, Basit M, Nisar MA, Khurshid M, Rasool MH. Proteomics: Technologies and their applications. J Chromatogr Sci 2017; 55(2): 182-96.
[http://dx.doi.org/10.1093/chromsci/bmw167] [PMID: 28087761]

[6]   Zumla A, Maeurer M. Rational development of adjunct immune-based therapies for drug-resistant tuberculosis: hypotheses and experimental designs. J Infect Dis 2012; 205 (Suppl. 2): S335-9.
[http://dx.doi.org/10.1093/infdis/jir881] [PMID: 22448017]

[7]   Hameed HMA, Islam MM, Chhotaray C, *et al.* Molecular targets related drug resistance mechanisms in MDR-, XDR-, and TDR-*Mycobacterium tuberculosis* strains. Front Cell Infect Microbiol 2018; 8: 114.
[http://dx.doi.org/10.3389/fcimb.2018.00114] [PMID: 29755957]

[8]   Bisht D, Sharma D, Sharma D, Singh R, Gupta VK. Recent insights into *Mycobacterium tuberculosis* through proteomics and implications for the clinic. Expert Rev Proteomics 2019; 16(5): 443-56.
[http://dx.doi.org/10.1080/14789450.2019.1608185] [PMID: 31032653]

[9]   Molecular dynamics simulation of antigen85 proteins (ag85a and ag85c) and an immunoinformatics study. 1995; 9-37. Available from: https://shodhganga.inflibnet.ac.in/jspui/bitstream/10603/196179/10/10_chapter%202.pdf

[10]   Jørstad MD, Marijani M, Dyrhol-Riise AM, Sviland L, Mustafa T. MPT64 antigen detection test improves routine diagnosis of extrapulmonary tuberculosis in a low-resource setting: A study from the tertiary care hospital in Zanzibar. PLoS One 2018; 13(5): e0196723.
[http://dx.doi.org/10.1371/journal.pone.0196723] [PMID: 29742144]

[11]   Zhu C, Liu J, Ling Y, *et al.* Correction to Evaluation of the clinical value of ELISA based on MPT64 antibody aptamer for serological diagnosis of pulmonary tuberculosis. BMC Infect Dis 2013; 13(1): 410.
[http://dx.doi.org/10.1186/1471-2334-13-430]

[12]   Singh V, Kaur C, Chaudhary VK, Rao KVS, Chatterjee S. *M. tuberculosis* Secretory Protein ESAT-6 Induces Metabolic Flux Perturbations to Drive Foamy Macrophage Differentiation. Sci Rep 2015; 5: 12906.
[http://dx.doi.org/10.1038/srep12906] [PMID: 26250836]

[13]   Sreejit G, Ahmed A, Parveen N, *et al.* The ESAT-6 protein of *Mycobacterium tuberculosis* interacts with beta-2-microglobulin (β2M) affecting antigen presentation function of macrophage. PLoS Pathog 2014; 10(10): e1004446.
[http://dx.doi.org/10.1371/journal.ppat.1004446] [PMID: 25356553]

[14]   Truong L, Harper SL, Tanguay RL. Evaluation of embryotoxicity using the zebrafish model. Methods Molecular biology. Methods Mol Biol 2011; 691(8): 1-8.

[15]   Kumar B, Sharma D, Sharma P, Katoch VM, Venkatesan K, Bisht D. Proteomic analysis of *Mycobacterium tuberculosis* isolates resistant to kanamycin and amikacin. J Proteomics 2013; 94: 68-77.
[http://dx.doi.org/10.1016/j.jprot.2013.08.025] [PMID: 24036035]

[16]   Sharma D, Kumar B, Lata M, *et al.* Comparative proteomic analysis of aminoglycosides resistant and susceptible *Mycobacterium tuberculosis* clinical isolates for exploring potential drug targets. PLoS One 2015; 10(10): e0139414.
[http://dx.doi.org/10.1371/journal.pone.0139414] [PMID: 26436944]

[17]   Sharma D, Lata M, Singh R, Deo N, Venkatesan K, Bisht D. Cytosolic proteome profiling of aminoglycosides resistant *Mycobacterium tuberculosis* clinical isolates using MALDI-TOF/MS. Front Microbiol 2016; 7: 1816.
[http://dx.doi.org/10.3389/fmicb.2016.01816] [PMID: 27895634]

[18]   Sharma D, Bisht D. Secretory proteome analysis of streptomycin-resistant *Mycobacterium tuberculosis* clinical isolates. Slas discovery: Advancing Life Sciences R&D 2017; 22(10): 1229-38.
[http://dx.doi.org/10.1177/2472555217698428]

[19]   Sharma D, Shankar H, Lata M, *et al.* Culture filtrate proteome analysis of aminoglycoside resistant clinical isolates of *Mycobacterium tuberculosis*. BMC Infect Dis 2014; 14 (Suppl. 3): 60.
[http://dx.doi.org/10.1186/1471-2334-14-S3-P60]

[20]   Sharma D, Bisht D. An efficient and rapid method for enrichment of lipophilic proteins from *Mycobacterium tuberculosis* H37Rv for two-dimensional gel electrophoresis. Electrophoresis 2016; 37(9): 1187-90.
[http://dx.doi.org/10.1002/elps.201600025] [PMID: 26935602]

[21]   Shouman W, El-Gammal M, Shaker A, *et al.* ESAT-6-ELISpot and interferon γ in the diagnosis of pleural tuberculosis. Egypt J Chest Dis Tuberc 2012; 61(3): 139-44.
[http://dx.doi.org/10.1016/j.ejcdt.2012.10.012]

[22]   Dewi DNSS, Mertaniasih NM, Soedarsono , *et al.* Characteristic profile of antibody responses to PPD, ESAT-6, and CFP-10 of *Mycobacterium tuberculosis* in pulmonary tuberculosis suspected cases in Surabaya, Indonesia. Braz J Infect Dis 2019; 23(4): 246-53.
[http://dx.doi.org/10.1016/j.bjid.2019.07.001] [PMID: 31421107]

[23]   Duraiyan J, Govindarajan R, Kaliyappan K, Palanisamy M. Applications of immunohistochemistry. J Pharm Bioallied Sci 2012; 4(6) (Suppl. 2): S307-9.
[http://dx.doi.org/10.4103/0975-7406.100281] [PMID: 23066277]

[24]   Solomon IH, Johncilla ME, Hornick JL, Milner DA Jr. The Utility of Immunohistochemistry in Mycobacterial Infection: A Proposal for Multimodality Testing. Am J Surg Pathol 2017; 41(10): 1364-70.
[http://dx.doi.org/10.1097/PAS.0000000000000925] [PMID: 28795999]

[25]   Kohli R, Punia RS, Kaushik R, Kundu R, Mohan H. Relative value of immunohistochemistry in detection of mycobacterial antigen in suspected cases of tuberculosis in tissue sections. Indian J Pathol Microbiol 2014; 57(4): 574-8.
[http://dx.doi.org/10.4103/0377-4929.142667] [PMID: 25308009]

[26]   Lavigne JP, Espinal P, Dunyach-Remy C, Messad N, Pantel A, Sotto A. Mass spectrometry: a revolution in clinical microbiology? Clin Chem Lab Med 2013; 51(2): 257-70.
[http://dx.doi.org/10.1515/cclm-2012-0291] [PMID: 23072853]

[27]   Colangelo CM, Williams KR. Isotope-coded affinity tags for protein quantification. Methods Mol Biol 2006; 328: 151-8.
[PMID: 16785647]

[28]   Sethuraman M, McComb ME, Heibeck T, Costello CE, Cohen RA. Isotope-coded affinity tag approach to identify and quantify oxidant-sensitive protein thiols. Mol Cell Proteomics 2004; 3(3): 273-8.
[http://dx.doi.org/10.1074/mcp.T300011-MCP200] [PMID: 14726493]

[29]   Shui W, Gilmore SA, Sheu L, Liu J, Keasling JD, Bertozzi CR. Quantitative proteomic profiling of host-pathogen interactions: the macrophage response to *Mycobacterium tuberculosis* lipids. J Proteome Res 2009; 8(1): 282-9.
[http://dx.doi.org/10.1021/pr800422e] [PMID: 19053526]

[30]   Büttner FM, Renner-Schneck M, Stehle T. X-ray crystallography and its impact on understanding bacterial cell wall remodeling processes. Int J Med Microbiol 2015; 305(2): 209-16.

[http://dx.doi.org/10.1016/j.ijmm.2014.12.018] [PMID: 25604506]

[31] Wong HC, Liu G, Zhang YM, Rock CO, Zheng J. The solution structure of acyl carrier protein from *Mycobacterium tuberculosis*. J Biol Chem 2002; 277(18): 15874-80.
[http://dx.doi.org/10.1074/jbc.M112300200] [PMID: 11825906]

[32] Rock CO, Cronan JE. *Escherichia coli* as a model for the regulation of dissociable (type II) fatty acid biosynthesis. Biochim Biophys Acta 1996; 1302(1): 1-16.
[http://dx.doi.org/10.1016/0005-2760(96)00056-2] [PMID: 8695652]

[33] Bortoluzzi A, Muskett FW, Waters LC, *et al. Mycobacterium tuberculosis* RNA polymerase-binding protein A (RbpA) and its interactions with sigma factors. J Biol Chem 2013; 288(20): 14438-50.
[http://dx.doi.org/10.1074/jbc.M113.459883] [PMID: 23548911]

[34] Sharma AK, Ye L, Baer CE, *et al.* Solution structure of the guanine nucleotide-binding STAS domain of SLC26-related SulP protein Rv1739c from *Mycobacterium tuberculosis*. J Biol Chem 2011; 286(10): 8534-44.
[http://dx.doi.org/10.1074/jbc.M110.165449] [PMID: 21190940]

[35] Stehle T, Sreeramulu S, Löhr F, *et al.* The apo-structure of the low molecular weight protein-tyrosine phosphatase A (MptpA) from *Mycobacterium tuberculosis* allows for better target-specific drug development. J Biol Chem 2012; 287(41): 34569-82.
[http://dx.doi.org/10.1074/jbc.M112.399261] [PMID: 22888002]

SUBJECT INDEX

0-9

2-DGE 138, 139, 140, 141, 142, 144
2D-PAGE 8, 19, 23, 25, 27, 30, 31, 104, 209

A

Acetylation 76, 77, 82, 83, 84, 85, 86, 88, 91, 95, 98, 99, 106, 206
Annotations 61, 62, 63, 66, 67
Antibiotics 1, 3, 4, 5, 6, 7, 10, 19, 79, 92, 93, 94, 158, 159, 160, 167, 172, 174, 191, 204, 206, 213

B

Bacterial Biofilm 1, 2, 3, 5, 10
Biomarkers 18, 19, 27, 28, 32, 40, 103, 147, 153, 154, 156, 157, 158, 159, 160, 161, 162, 179, 180, 181, 182, 184, 195, 204, 211, 213, 214

C

Carboxylation 77, 88, 96, 102

D

DeCyder 18, 26, 27, 28, 29, 30
Detergents 19, 20, 138, 139, 140
Diagnosis 37, 61, 65, 148, 153, 157, 158, 159, 160, 161,184, 204, 205, 207, 208, 210
Diagnostics 147, 157, 158, 160, 204, 207, 210
DIGE 18, 19, 26, 27, 28, 29, 30, 31, 104, 156,157, 180
Drug discovery 67, 180, 192, 193,
Drug resistance 1, 2, 4, 5, 6, 7, 10, 79, 92, 93, 94, 103, 131, 167, 172, 182, 183, 204, 205
Drug targets 79, 100, 157, 180, 181, 182, 191, 193, 194, 195, 204, 214

E

Electrospray 36, 65, 158, 169, 211

G

Gel electrophoresis 8, 18, 19, 20, 21, 23, 24, 25, 26, 27, 28, 29, 38, 42, 46, 65, 82, 138, 156, 161, 180, 181, 182, 205, 208, 209
Glycosylation 19, 76, 77, 80, 82, 84, 85, 88, 90, 91, 96, 99, 102

H

Host-pathogen interaction 78, 85, 87, 88, 148, 149, 167, 173, 187, 194, 205

I

ICP-MS 167, 169, 170,
IEF 18, 20, 23, 24, 25, 30, 31, 139, 140
Infectious diseases 77, 79, 147, 149, 150, 153, 161, 162, 173, 174, 185
Isobaric Tags for Relative and Absolute Quantification (iTRAQ) 9, 40, 66, 106, 190, 191, 212
Isotope-coded Affinity Tags (ICAT) 8, 39, 66, 104, 211

L

Lipidation 77, 88, 92, 95, 99, 101
Liquid Chromatography 9, 19, 36, 38, 42, 66, 154, 158, 161, 179, 180, 189,206, 213
Liquid Chromatography Mass Spectrometry (LCMS) 36, 38, 48, 50, 52, 180, 189

Divakar Sharma (Ed.)
All rights reserved-© 2020 Bentham Science Publishers

M

M. tuberculosis 95, 96, 98, 100, 103, 179, 180, 182, 183, 184, 185, 186, 187, 188, 189, 190, 191, 192, 193, 194, 195
Mass spectrometry 18, 19, 27, 38, 40, 41, 46, 65, 66, 82, 85, 96, 104, 107, 126, 140, 147, 149, 154, 156, 158, 159, 161, 170, 180, 190, 204, 209, 211
Metabolic pathways 52, 180, 183, 209
Metalloproteome 167, 174, 175
Metalloproteomics 167, 168, 170, 171, 174, 175
Methylation 76, 77, 84, 85, 86, 88, 94, 95, 96, 97, 206
Microbial drug resistance 162, 172
Microbial proteins 32, 61, 62, 66, 77, 138, 141
Microbial proteomics 18, 22, 31, 32, 36, 37, 54
Microorganisms 1, 31, 36, 43, 52, 62, 63, 77, 79, 95, 159, 173
Multidrug-resistant TB (MDR-TB) 180, 183
Mycobacterial proteasomal ATPase (Mpa) 122, 123, 124, 125, 126, 130, 131

N

Nitrosylation 77, 88, 96, 102

P

Pathogenesis 1, 6, 32, 76, 84, 89, 94, 95, 101, 122, 147, 148, 149, 155, 157, 158, 167, 169, 191
Phosphorylation 19, 42, 76, 77, 82, 83, 84, 85, 88, 89, 90, 95, 98, 99, 106, 148, 168, 180, 181, 183, 194
Posttranslational modifications (PTMs) 37, 42, 43, 76, 77, 78, 79, 80, 81, 82, 83, 84, 85, 87, 88, 89, 90, 91, 92, 93, 94, 98, 99, 103, 104, 105, 107, 148, 212
Proteasome system 86, 99, 129, 130, 131, 132
Protein sample preparation 19, 138, 139, 144
Proteomic tools 77, 79, 82, 104, 181, 195

Pupylation 91, 92, 96, 99, 103, 122, 123, 126, 127, 128, 129, 130, 131

S

SDS-PAGE 7, 8, 18, 19, 21, 22, 23, 25, 30, 31, 208
Stable Isotope Labeling by Amino Acids in Cell Culture (SILAC) 8, 39, 44, 104, 106, 152, 190, 212
Structural biology 180, 191, 192, 193
Structural genomics 180, 191, 192
Systems biology 61, 62, 67, 68, 72, 73, 174

T

Therapeutics 92, 100, 147, 148, 157, 158, 167, 179, 204, 214
Transcriptomics 2, 10, 36, 61, 149, 155, 208

V

Vaccine 52, 53, 147, 155, 156, 158, 161, 180, 182, 183, 185, 186, 187, 188, 204, 205, 207

X

X-ray absorption 167, 171, 172

CBI m/pod-product-compliance
4800